Hooks for the Fly

0 11557 02803 4

Hooks for the Fly

William E. Schmidt

STACKPOLE
BOOKS

Published by
STACKPOLE BOOKS
5067 Ritter Road
Mechanicsburg, PA 17055
www.stackpolebooks.com

Printed in the United States

10 9 8 7 6 5 4 3 2 1

First edition

Cover design by Wendy A. Reynolds

Library of Congress Cataloging-in-Publication Data
Schmidt, Bill (William E.)
Hooks for the fly / William E. Schmidt.— 1st ed.
p. cm.
Originally published: Buzzards Bay, MA : Piscator Pub., © 1997
Includes bibliographic references (p.).
ISBN 0-8117-2803-X (pbk.)
1. Fly tying—Equipment and supplies. 2. Fly tying—Equipment and supplies—Directories. 3. Fishhooks. I. Title.

SH451.S315 2000
688.7'9124—dc21

00-029697

To Lisa, Drew, and Caleb. May you find joy as I have found in you.
And to those who have the fly-fishing passion as I do.

CONTENTS

Acknowledgments	viii
Preface	ix
About Hooks	xi
1. Daiichi	1
2. Dai-Riki	14
3. Eagle Claw	17
4. Gaelic Supreme	24
5. Gamakatsu	29
6. Hayabusa	34
7. Heritage	38
8. Kamasan	41
9. Mustad	47
10. Orvis	62
11. Owner	67
12. Partridge	69
13. Tiemco (TMC)	88
14. Turrall	97
15. Varivas	102
16. VMC	104
Appendix	
Specialty Hooks	111
Manufacturer Contact List	117
References	119

ACKNOWLEDGMENTS

To start, I'd like to thank the hook manufacturers themselves: Daiichi and Bill Chase of Angler Sport Group; Dai-Riki and the people at Dan Bailey's Fly Shop; the Wright and McGill Company, specifically George Large; Glenn Young at Gamakatsu; Skip Mortensen, who gave this project a big first boost, and on his relief, an array of great people at O. Mustad and Sons, USA; Tom Rosenbauer and Kevin Sloan at Orvis; Alan Bramley, who shared with me the history of the company, and several others across the big pond at Partridge; Grahame Maisey at Belvoirdale, the supplier of Gaelic Supreme hooks, with whom I enjoyed long talks; Jim Watson, Bruse Olsen, Ken Menard, and all the other people at Umpqua Feather Merchants; Steve Pranskey of Turrall, USA, and Tod Marshman of H. Turrall; and Zack Swanson at VMC. Someday soon I will visit your factories and put faces to all those voices and letters. Keep up the good work. One more thing: Thanks to all of you for convincing my wife this was a worthwhile project.

For the self-published first edition, I thank my three editors. Tony Palmer, my fellow Coastie and long-distance fly-fishing partner, thank you for really stoking my fly-fishing desire. We first fished together in Alaska, and now we fish in New England; let's make it the rest of the world when we retire. The people who attend your Palmer Fly Fishing Clinics are lucky. Chris Helm, author, video producer, and proprietor of Whitetail Fly Tieing Supplies, your support, input, and knowledge helped more than you imagine. I wish I could have known you when you taught at my high school. Mike McNutt, you have been a mentor to me throughout the years as both a musician and fly fisherman. I can't wait until I can go fishing with all of you again.

Ken Ferguson at Spirit River, Inc., you're right about the Big Game hook. It's become a staple in my salt tying arsenal. Thank you.

Joe Dion, a very knowledgeable person on hooks as well as an avid fly fisherman, you were a big help in establishing these cross-reference and characteristic tables.

Art Scheck, editor of *Fly Tyer* magazine, your ideas and tolerance of my pesky questions were noble. You and your staff have become mentors for our sport. I hope to visit with you again, but this time to fish.

A warm, heartfelt thanks to Lefty Kreh. I was very nervous about showing you the first edition. Your kindness put me at ease right away, and the words you wrote about the book made me feel much more proud of the work. I was already a fan of yours but am now an avid admirer and devout student of your teachings.

Mike Martinek, veteran fly tier, collector of fly-tying items, and expert on New England streamer flies, your variation of the Carrie Stevens style streamer hook is well thought of and will be in my repertoire of hooks for many years to come.

I can't forget to mention Plaza Sports in Ketchikan, Alaska. You let me run with the fly-fishing section of your fine store. You taught me a lot. Keep in touch, Jack.

Heartfelt thanks and a big round of applause to the folks at Stackpole Books, specifically to Judith Schnell, for taking a chance with my work, and to Jon Rounds. Jon, your patience and enthusiasm were a joy to work with.

To my mom and dad, Bill and Gayle: Your faith in me and all my projects and trials is undying. I enjoy making you proud. Thank you for the support and love you've given me throughout my years.

And finally, to the people who really had to deal with me day in and day out—my wife, Lisa, and sons, Drew and Caleb—you are my future and the loves of my life. Lisa put many hours of work into this edition. She kept me focused and humble as she learned about hooks. If she doesn't know anything about hooks, no one does! Drew was once allergic to fur and feathers, but as the last three years passed, he got stuck with needles every week. I am happy he can now join me at the bench. Caleb, being five, was too young to be at the bench, but not too young to start training for using the results at the bench—let's go fishing!

PREFACE

Inspiration for this book was born out of frustration. You see, at best, I'm a novice fly tier and an artist only in my own mind. How many times have I searched a fly-tying book or magazine for a new fly, only to discover I don't have the hook called for? The fly is bastardized from the start. Never mind my lack of skill, my hands shaky from caffeine, my eyes that need three lights and a magnifier. Finding the hook becomes an added task that consumes precious time.

I was taught—and still live by—a few rules about fly-fishing and fly-tying equipment: Always use the best waders, fly lines, leader material, and hooks. Always get the best vise you can afford, keep your scissors sharp and in your hand, and use the brightest light available. As for the rest of the equipment, buy the best you can afford.

The question of which is the best brand of hooks has to be answered by each fly tier. There are more than sixteen manufacturers and distributors of fly hooks, and most have been in business for a very long time. This suggests that many fly tiers are loyal to a particular brand. You must actually try a hook to gain confidence in it. It must stay whole in the vise, have the right profile for the imitation you're creating, and pierce and hold a fish well. It should let you concentrate on the task at hand instead of wondering whether it will fail on your fish of a lifetime.

From its very inception, this book began to help me. It started with some notes comparing information in catalogs and brochures, and expanded to information I obtained on the Internet and then from talking to manufacturers themselves.

I started writing this book while I lived in Ketchikan, Alaska. I only had access to hooks with packaging that said, "Dry Fly," "Nymph," or "Streamer," a brand of hooks I'd heard of but never used. Oh sure, I could order from many catalogs. (I have a collection that can fill a two-drawer filing cabinet.) But what could I do until the box of $6 hooks arrived? (Overnight air express cost an additional $16.) My recipe called for a Tiemco, Partridge, or Mustad, and I didn't have any left. You can run down to that open-twenty-four-hours, got-everything-under-the-sun sporting-goods shop and be back tying your fly inside twenty minutes. I, in the meantime, had to explain to my wife why I had to drive 15 miles one-way into town and miss dinner to pick up a ten-pack of hooks.

Eventually I worked part-time at one of the two sporting-goods shops in town, in the place the townsfolk called the mall. (It was the biggest building in town, other than the cruise ship tied to the pier.) I was hired because I knew something about camping and fly fishing. The fly-tying section took up a whole end cap of one aisle. The other shop didn't do fly tying, so this end cap was basically the only game in town. I knew quite a few people in town tied flies, and they did what I did—ordered from catalogs. The manager and I built up the fly-tying section until it eventually grew to a respectable one side of an aisle!

Now I had easy access to hooks I had never used before, but I still didn't know which hooks corresponded to the ones my recipes recommended. So I browsed on the Internet for a while and made a cross-reference table for myself. Then I got this brilliant idea to help all of my fly-tying buddies out there—cross-reference *all* the fly hooks! (More likely, I was bored with the sun setting at two in the afternoon and rising at nine in the morning.) After a year and a half, a job transfer from Alaska to Cape Cod, and a computer crash or two, I had a book. Now, three years after that, two casting arm surgeries (learned the hard way I was casting wrong), and another job transfer, I present to you this edition of *Hooks for the Fly.*

All my original tables, the small tables downloaded from the Internet, and those given to me by private citizens were filed away when I decided to go public with these tables. I started from scratch. I researched every catalog, brochure, flyer, and ad slick put out by the manufacturers. I tried my best to rid any biases from these tables.

Many tiers wrote to me saying this book is what they've been looking for, and some made great suggestions. Most, if not all, of those suggestions are included here. The largest change came from requests for pictures of the hooks and size listings. During my tying sessions since the first edition, I, too, discovered a need for such information.

Some hooks are sold under their manufacturer's name, while others are made by these same manufacturers but distributed under a different name. I've decided to refer to both manufacturers and distributors as "makers," because the manufacturers make the hooks and the distributors make the hooks available to us.

Enjoy your fly tying, and may you always get a rise to your fly.

ABOUT HOOKS

HOOK DESCRIPTIONS

All hooks are described using similar terminology and acronyms. For example, 1XF, 2XL, TDE describes a hook of size 1 extra fine wire, size 2 extra long shank, and with a turned-down eye. The List of Abbreviations on pages xvii–xviii lists all the descriptors in use today. This list consolidates any minor variations of the descriptors. For example, the wire descriptor for larger wire is either H for heavy or S for strong or stout, depending on the maker (in other words, 1XH = 1XS wire). To avoid confusion, this book uses an H, because using an S as a wire descriptor can be confused with the shank descriptor S, which stands for short. Not only that, but S alone means standard. Which brings us to a very sticky subject. (Confused yet?)

STANDARDIZATION

Standardization is sadly lacking in the hook industry. Each maker has its own standards, for the most part anyway, but none exist from maker to maker. We tiers have complained about this for years, but our complaints have fallen on deaf ears. We have not been able to force hook makers to commit to standardization as the fly-line industry has. Hook makers have been around for too many years and are very set in their ways. Yet this trait has allowed tiers to have a wide choice of hook styles. I don't know about you, but I would rather have choices than standards. It adds to the fun of fly tying. Adaptation and experimentation are at the very heart of our art.

This book's main goal is to help you determine which fly hooks are *similar by characteristic descriptions.* This will allow you to judge which hook best fits your pattern. One or 2 millimeters or milligrams is not a big difference for most patterns, although it may matter when tying match-the-hatch patterns. Then again, this difference can be area-specific. An originator of a pattern or author of a pattern book or article may have reasons other than the fly pattern's needs to use a particular hook. If your recipe tells you to use a Mustad 3906 for a Gold-Ribbed Hare's Ear nymph, use it if you have it. But if you can't get that hook or prefer another brand, there is no reason not to use a Daiichi 1550 or a Turrall Sproat instead. The pattern will not be degraded. However, if you do want or need to be exacting to match your area's hatch, you now have a list of similar hooks.

Among makers, there are similarities of hooks in different categories. For example, one hook is listed as a standard by one maker and as a 1XL by another, but there is no visible difference. Most of the makers have discovered such discrepancies and incorporated them into their cross-reference tables, and this book has compiled these.

If you are looking for specific hook characteristics, and a hook doesn't quite match, I suggest you "bracket" the hook. That is, find the hooks that are one step larger and one step smaller. You may find your desired hook listed as something other than what you think it is. Also, a change in a hook size from another maker may do the trick. Use this book to help you narrow your search and decide what is best for the fly you're tying, designing, or adapting to your waters.

Among makers, no two hooks are exactly the same (unless the same hook is distributed under another name). The length may be the same but the weight will be different, or vice versa. If, by remote chance, two hooks are exactly the same in length and weight, they still won't be the same. Maker 1 cannot create exactly what Maker 2 has, unless Maker 1 has stolen or taken a very lucky guess at Maker 2's metallurgical formula, the trade secrets that make or break a company. Then again, the two hooks may be close enough in this regard to what most of us need.

Some standards do extend among makers. They involve what is done to the wire after the weight and length have been determined, and include hook parts, bend styles, eye shapes and positions, and finish application and color.

SIZES

Scales used to determine the size of a hook in a particular style aren't fully employed across the industry, and those used for one model can't always be applied to another. It's difficult even to define hook size. Some say that size is a measurement of the gape. That seems simple enough until you try to size hooks with bent shanks. There are several hook scales, some published and some obsolete, including the Standard Scale, Redditch Scale, Mustad's Viking Scale and handy scales you can attach to your vise. These scales work in some instances, sometimes specifically for hooks by one maker only. Some scales consider the length of the hook only, and disregard the gape; others look at the gape and not the length. The inclusion of hook profiles in this book should help you determine the size of a particular hook.

One hook model usually comes in several sizes. A model is defined by the overall style, based on the hook's shape and desired use, and characteristics of the hook, including the wire size, shank length, bend type, eye position, gape, throat depth, point type, and barb. These components are proportionately consistent throughout the model's sizes, although changes in wire size or gape, for example, may be required.

Sizes are usually given in even numbers, although a few models are odd numbered. (Odd-numbered sizes are slightly different in overall size from the closest even-numbered hook.) The smaller hooks run from size 2 to size 28. The smaller the hook, the larger the number. The size labeling system changes for larger hooks. The smallest of the large hooks is 1/0, sometimes pronounced "one aught," and they are labeled consecutively—2/0, 3/0, 4/0—as the hook size increases. In between the smaller and the larger hooks is size 1. Most makers who make the larger-size hooks in a particular style have a size 1 hook as well.

HOOK COMPONENTS

Bend

Manufacturing the metal hook's bend the old way, by physically bending the wire by hand around a mold, is still done today. Handmade hooks are still produced or marketed by Partridge, Turrall, Gaelic Supreme, and Heritage. The molds are made from metal dies, carved out of wood, or have dowels placed precisely in the shape of the desired bend. An interesting book, *The Colonial Angler's Manual of Flyfishing and Flytying,* by Ken Reinard, explains how to produce your own hooks using the colonial method.

Bend shapes and styles tend to be standard throughout the industry. No single bend shape used today stands out more than any other. Each has its pros and cons, and there seems to be little difference in hooking and holding ability. Some industry people say that a Limerick bend offers better penetration because of its straight spear. But hooks with Perfect and Aberdeen bends give good penetration also. Any bend offered today will catch and hold fish. This is especially true when the hook is used in the fishing conditions it was specifically designed for. The most notable differences between bends are purely esthetic. If you want your fly to look its best for show, then the bend comes into play. One would have a tough time convincing me a fish fails to hit a well-dressed hook because of the bend shape. It's the stuff on the hook, not the hook itself, that's supposed to attract the fish, although some of the new hook finishes offered today are changing this view.

HOOK COMPONENTS

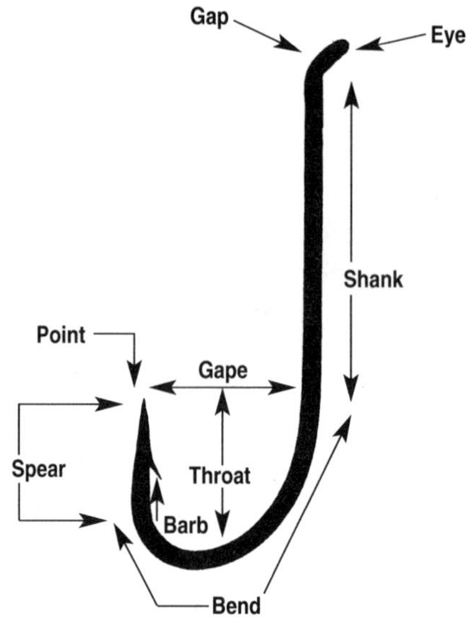

Some bends are better for certain patterns. The profile and proportions of a fly stem from the hook itself. Proper tail length is almost always determined by the length of the shank, and the height of the wings is usually measured against the gape. Buoyancy or lack thereof is a large consideration that depends on the fly category. Dry flies are usually tied on light hooks, and nymphs or wet flies on heavier ones. Some floating fly designs, such as those tied with deer or elk hair, require a stout hook. A light hook

Bends and Styles

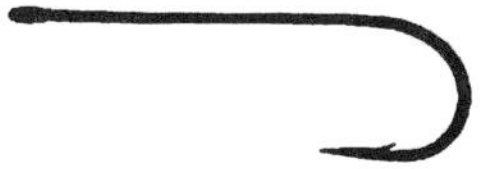

Aberdeen
Some consider this a hook style with a Perfect bend

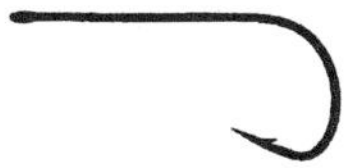

Sneck
Also called Wide Gap
Usually a "stinger" style

Sproat

Perfect
Similar bends: Model Perfect, Round, Hamilton, and Viking

Dublin
Very similar to a Bartleet as well

O'Shaughnessy

Beak
Sometimes called Octopus Hook

Wilson

Limerick

York

Continuous
Sometimes called Wide Gap

Offset Bends

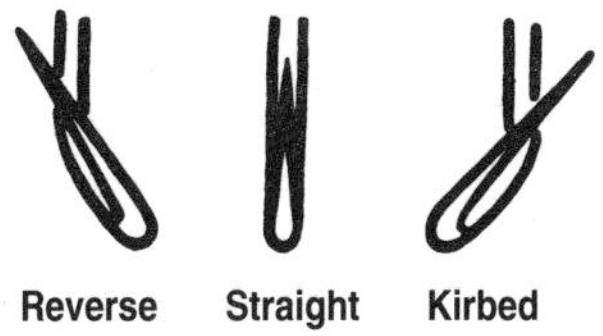

with the right style of components can also work for a wet fly, and a heavily hackled fly may sit on the water better with a heavy hook chassis.

Hooks can have offset bends, known as reversed or kirbed, depending on which way the point sticks out from the shank. If you hold the hook eye up, with the point toward you, and view the point, it may be aligned with the shank (not offset), offset to the right (kirbed), or offset to the left (reversed). Hooks offset with bends are not widely used in fly tying. Offsetting may increase the probability of a hookup, but it's a little more difficult to tie materials to such a hook.

Eye

Fly hook eyes can be normal sized, oversized, and undersized and come in three types: ball or ringed, looped, and blind. Blind hooks are those having no eye. The other two have been modified further, into tapered ball and tapered looped. The tapering applies not only to the section of wire the eye uses, but also to a length of the shank directly behind the eye. Tapering helps make the hook lighter. Looped eyes, tapered and not tapered, are mainly used for salmon and steelhead flies.

Not all hooks are made perfectly. Inspect the eyes of the hooks to ensure that the gap between the bitter end of the eye is as close as it can be to the shank. An exception is if the hook is an open eye hook and the gap was designed to be there.

Eye positions can vary. The eye can be turned up, turned down, straight, partially turned, or flat/vertical. A ringed eye can refer to a flat/vertical eye, or straight eye, depending on the maker. Generally, a turned eyed hook is one that is bent to a 45-degree angle to the shank. Several hooks have modified eye turns to 20 degrees. The general consensus is that a straight-eyed hook has better hooking ability but does not

Eye Positions

Down

Up

Straight or ringed

Vertical or flat

Eye Types

Ball or ringed

Tapered ball

Looped
(may be round or triangular)

Tapered loop
(may be round or triangular)

Open eyed
(used in any position)

always allow a desirable fly riding position. Up-eyed hooks are said to allow the fly to ride better.

Differences in eye size and tapering are not criteria used in cross-referencing hooks in this book. The position and shape of the eyes seem to be standard throughout the industry, although variations do exist.

Barb

In the last few years, all my freshwater fishing has been with barbless hooks. I truly believe I haven't lost any more fish than I would have if I had used hooks with barbs. Barbs damage fish's mouths, are a pain to get out of the drying fleece, and are hazardous to us dummies who hook our ears while false casting.

If you don't want to pay extra for barbless hooks, which cost more than the standard barbed models, you can smash the barbs or file them down. Smashing them leaves a hump, which may provide a little extra holding power. Micro- or minibarbs are quickly becoming the norm for newly designed hooks. This is a good modification. The minibarb still does the job a barb is designed to do but with much less damage to the fish. The minibarb is smaller and therefore allows quicker penetration and is easier to smash. Hook technology is ever advancing, and we may someday see the barbless hook as the standard. The barb, or lack thereof, is not a factor in the cross-reference tables.

Point

Many point shapes are on the market today, all of which have their pros and cons. These include cone shapes, triple edges, knife edges, rolled, and many more.

Having a sharp hook point is essential and humane. It's easier to sharpen a hook before tying any materials on it. Carrying a sharpening stone or steel to the water is also prudent. I would estimate that more than 95 percent of modern hooks are sharp enough to be used straight out of the package. Stainless steel hooks can be sharpened all you want without fear of rapid corrosion, but if your hook has a finish, don't sharpen it if you can stick the point in your thumbnail and it's obviously very sharp. The bronzing, japanning, tinning, or other finish is there to deter rusting. If you remove the finish from the point of the hook, you're allowing corrosion to enter the thinnest part of the hook faster. As soon as that hook hits the water or is exposed to moisture in your fly box, corrosion starts. By the end of the day, your point will be duller and weaker.

Chemical sharpening takes place after the hook has been physically sharpened and before the finish is applied. The hook is placed in a chemical bath to remove the tiny barbs and ridges always left behind from the sharpening process.

One hook and point style that's a little different is the tag hook. A tag hook has a blunt point or an eye where the point ordinarily would be. You tie your materials on this hook and fish with it the same as any other. What is significant about using tag hooks is you do not hook up with the fish. The fish takes the fly, you simulate the hook set, the fish tugs, you tug, and the fish lets go none the wiser. I think we could use these hooks to open the rapidly increasing areas closed to fishing. The fish is not mortally injured by this method. You get to enjoy the area and the fishing with little impact.

Shank

Shanks are available in many different lengths and shapes. They are ordinarily measured from the base of the eye to the beginning of the bend, although in continuous bend hooks, the shank can't be measured this way. Generally, continuous bend hooks are classified as standard length shank. You can get hooks from very short (7XS) to very long (10XL) and everything in between. Shapes are available in straight, sweeping curves, bent down, bent up, keel bends, curved up (swimming nymph), with one or two popper holding kinks, and other special designs. Mustad has a few hooks that are described as S+$^{1}/_{2}$", S+$^{1}/_{4}$", and S+ $^{3}/_{8}$". This indicates that a shank is the standard length plus the given fraction. The S+$^{1}/_{2}$" is considered equal to a 6XL.

Some specialty hooks produced by Partridge are interesting in design. Some have shanks that are split lengthwise, creating a draper-style hook to help make a flat nymph body. Another hook looks like a naked mayfly.

Spear

The spear encompasses the part of the hook from the point and barb to the bottom of the bend. This is the part that you drive into the fish.

Special Shanks

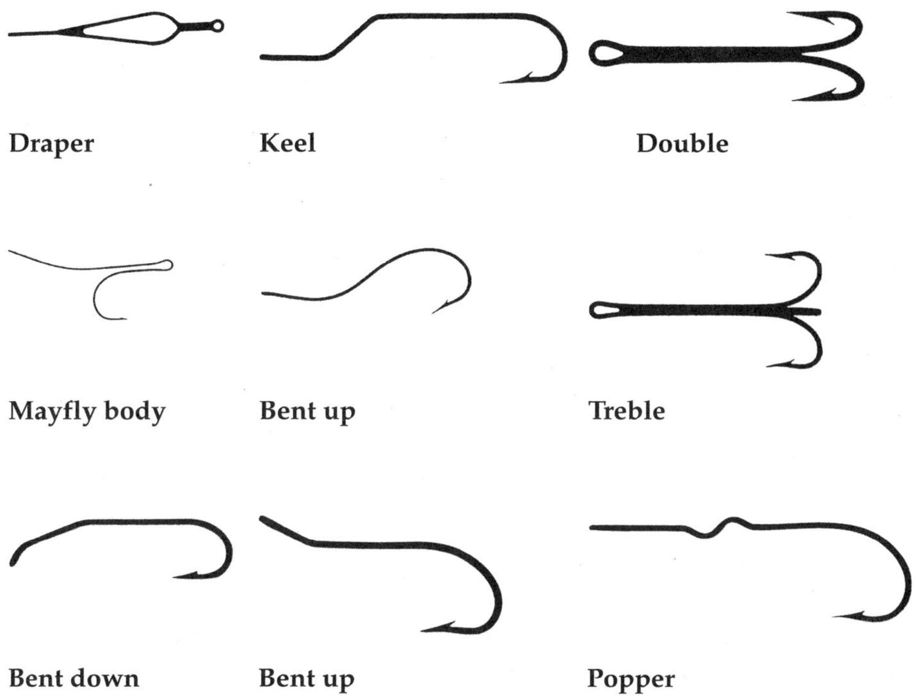

Throat

The throat is the "holding area" of the hook—the area between the shank and the spear from the deepest part of the bend to the point. The smaller the throat, the more skill you need to play the fish.

Gape

The gape, refers to the distance between the point tip and the shank.

WIRE

The wire of some hooks is made from high-carbon steel, which has produced some very strong hooks with small-diameter wire. The wire is proportioned up or down, based on the size in a particular model. Some hook models use a different size wire for their smaller hooks. This allows the larger hook sizes to be lightweight and the smaller sizes to be just as strong as the larger hooks. In the characteristic tables, I have identified which hooks make such a change. The range of wire runs from 6XF to 1XF, Standard, and 1XH to 5XH. (The letters indicate extra fine or extra heavy.) The higher the number, the lighter or the heavier the hook.

FINISHES

Various makers have developed finishes that make carbon steel hooks more corrosion resistant. The name of the finish and shades of color vary from maker to maker. Hooks with such a finish will corrode more slowly, and thus you should consider them as your replacements for stainless steel hooks. Although stainless steel hooks are fairly corrosion resistant, it takes many months or possibly years for a stainless steel hook a fish breaks off to corrode and fall out of the fish. Use bronzed or japanned finish hooks in fresh water and tinned hooks in salt wherever possible.

THE TABLES

The hooks in the three sets of tables—cross-reference, characteristic, and specialty—are grouped by maker.

CROSS-REFERENCE TABLES

The cross-reference tables are set with the hooks of the primary maker in the far-left column. The remaining column titles show the other makers that have hooks that are similar in characteristic description, in alphabetical order. When looking for a particular hook, find the table containing the primary maker's hook that you want, then locate the cross-reference maker you're comparing it to. Read across from the primary hook called for to find a cross-referenced hook. Makers not included in these tables are those that offer no hooks cross-referenced to the primary maker.

The footnotes on individual hooks appear on the last page of each maker's cross-reference table. Where there are visible differences in bends, finishes, or points, these are noted to help you determine what type you need for a particular fly pattern. A hook with characteristics 1XF, S, TD may be available with a Perfect bend by maker 1, and with a Sproat by maker 2. You decide which bend best fits your pattern. Or maybe your pattern calls for a hook with a Perfect bend by maker 1 but you only have hooks from maker 2. The hook you have is the same except for the bend, not a very big difference in most instances. Just ensure that the fly pattern can tolerate the slope of the Sproat bend before you use the cross-referenced hook.

Hooks located in the columns to the right of the primary maker's column may or may not cross-reference to each other. Let's say you're looking at the Daiichi table, and there are eight hooks from three other makers, Mustad, TMC, and Dai-Riki, that cross-reference to the Daiichi hook. Although the Mustad, TMC, and Dai-Riki hooks cross-reference to the Daiichi, they may not cross-reference to each other. The Daiichi hook may be the middle of the road for that particular style, with the Mustad hook at one end of the style and the TMC at the other. The Mustad and TMC hook may not be similar enough to replace each other. This is why there are cross-reference tables for each maker.

Characteristically similar barbed and barbless hooks are grouped together. ***Bold italicized*** hooks are those that are barbless. Other than this, barbless hooks are not differentiated from the barbed hooks. Also, eye size and tapering are not cross-referencing factors and thus are only identified in the characteristic descriptions.

Some hooks have recently been discontinued. These hooks are listed in their makers' primary cross-reference tables but do not appear in any other maker's cross-reference table. There's no sense in cross-referencing you to a discontinued hook that is no longer available.

CHARACTERISTIC TABLES

The characteristic tables are what the cross-reference tables are based upon. Use these tables to make your own comparisons and to discover hooks that you may not have been aware of. Regretfully, these tables are loaded with abbreviations that were unavoidable due to space constraints.

These tables are in alphabetical order by maker. The column labeled Chief Use lists the main advertised use. Other advertised uses or suggested use possibilities are in the Remarks column. These uses are not limitations; your experimentation with hooks may lead you to discover some new applications. Who knows, maybe you will be the next inventor of a great fly or fly style.

The sizes are listed from the largest size in which the hook is currently available to the smallest.

The sample hooks images were taken directly from what the makers have developed. Because many makers do not provide images to scale, some sizes of sample hooks are not listed. If you identify any inaccuracies, they stem from the original images provided to me.

So break out your jar of loose hooks, scrape off the ugly tied flies (everybody has them) and salvage the hooks, identify what hooks they are, and use them again.

QUICK-REFERENCE SPECIALTY HOOK LISTS

These lists are provided in the appendix to differentiate hooks with special characteristics or those that are advertised as having an application other than the traditional uses. As an example, saltwater hooks are not necessarily only made of stainless steel, and stainless-steel hooks aren't only for salt water, so there's a list here for each. There are lists of manufactured barbless hooks, as well as bass bug, popper, tag, and blind eye hooks. Other lists are of hooks with additional components to make special bodies, hooks with bent-up or bent-down shanks, including the curved swimming nymph hooks, and hooks in colors other than bronze, black, or variations of silver.

ABBREVIATIONS

Wire

S	Standard wire diameter/weight for that maker's style/ model of hook.
1XF	One unit size finer (lighter) than standard. Continues 2XF, 3XF and so forth.
1XH	One unit heavier (stronger) than standard. Continues 2XH, 3XH and so forth.
(Rmks)	Read the remarks section. This usually means the wire size for the hook changes for the size of the hook.

Shank

S	Standard length for that maker's style/model of hook.
1XL	One unit size longer than standard. Continues 2XL, 3XL up to 10XL.
1XS	One unit size shorter than standard. Continues 2XS, 3XS and so forth.
S+1/2"	One half inch longer than standard. Equal to a 6XL.
S+1/4"	One quarter inch longer than standard.
S+3/8"	Three eighths of an inch longer than standard.
-BU	Bent up. Special purpose hook.
-BD	Bent down. Special purpose hook.
-C	Curved
-K	Keel bend
-PK	Popper kink(s)
-S	Special design. See Remarks.

Eye

TD	Turned down
TU	Turned up
S	Straight, ball, or ringed
F	Flat or vertical
-L	Looped
-M	Modified (not fully turned)
-O	Oversized
-U	Undersized or needle eye
No	No eye or blind eye

Bend

Aber	Aberdeen
Beak	Beak, octopus
C	Continuous
Dlim	Dublin-Limerick
Dub	Dublin
Lim	Limerick
N.I.	Bend not identified
Osh	O'Shaughnessy
P	Perfect, Model Perfect, Round, Hamilton or Viking
Sneck	Sneck, stinger
Spr	Sproat
WG	Wide gape
Wil	Wilson
York	York
-Dbl	Double bends
-M	Similar but modified or improved in some way
-S	Special (unique to each maker)
-R	Reversed or Kirbed (point bent out to side)
TREB	Treble (three) bends

Finish

Blk	Black, japanned or enameled
Blue	Blue
Bron	Bronzed
Corr-*[color]*	Special corrosion resistant. See Remarks.
Gold	Gold
Grn	Green
GS	Gray shadow
M-Nick	Matte nickel
Nick	Nickel
Olive	Olive enameled
Red	Red
(Rmks)	See Remarks column for specifics
Sil	Silver
SS	Stainless steel
Tef	Teflon coated
Tin	Tin plated or tin hook with other finish

Barb

No Barb	Barbless
Mini	Mini or micro barb
Out	Barbs are outside of the throat

Point

Chem	Chemically sharpened
-C	Chemically sharpened
DP	Dropped point
Out	Out point
SDP	Semi-dropped point
Spec	Maker's special design. See Remarks.
Tag	Dull or eyed point

Remarks

SJW	San Juan Worm
Mfg	Manufacturer

Daiichi	Dai-Riki	Eagle Claw	Gaelic Supreme	Gamakatsu	Hayabusa	Kamasan	Mustad	Orvis	Owner	Partridge	TMC	Turrall	Varivas	VMC
1100		L059 L060[3] ***L061B***		S10-3F[2,4]	572[3]	B401[2,3] B400[3]	94840 AC94840 ***94845*** 80000BR ***80010BR*** 94834[4]	4864[1] 1523		L3A[3] ***L3AY***[3] GRS3A[3,4]	5210 100[1,3] ***100BL***[3] 146Z[4] 102Y[3,4] ***103BL***[3,4] ***900BL***[3,4,5]			9280[3] V7061[2]
1110				S10S[2,3] S-13S-M[4] S12S-1F[3]			94859	4641[1,3] 170T[3,4]			101[1,3]			
1130	135[2]	L055[1,2,3]		C12[2,3]			80250BR ***80260BR***	1639[1,2,3]		K4A[1]	2487[1,3] ***2487BL***[3] 2487G[3]			
1140							80100BR[1,3]	1639[1,2]			***206BL***[1,3]			
1150	135[2]				371[3]	B420[2,3]	80200BR[1,3]	8891[3]		K2B[3]	2457[1]			
1155					371[3,4]	B420[3,4]		8891[3,4]		K2B[3,4]	2457[1,4]			
1170	300[1,2] 305	L059[1]		P10[2] S10[2,4] ***S10-B***[4]	372 373[3] ***378*** 386	***B402*** K40 B401[3]	94840[1]	1876 1509[3] ***1877***[5] 4864[3]		L2A[1,3]	5210[1]		980	9280[1,3] 9060 8527[3]
1180	305	L059[2]		P10 S10[4] ***S10-B***[4]	372 ***378*** 386 373[3]	***B402*** K40 B401[3]	94840[1]	1523[1,2] 1509[3] ***1877***[5] 4864[3]			5210[1]		980	9060
1182	305[4]			P10[4] S10[4] ***S10-B***[4]	372[4] 373[3,4] ***378***[4] 386[4]	***B402***[4] K40[4] B401[3,4]		1509[3,4] ***1877***[4,5] 4864[3,4]					980[4]	9060[4]
1190	305	***L061B***[1,2]		***S10-B***[2,4] P10[4] S10[4]	372 373 ***378*** 386	***B402*** K40 B401[3]	***80010BR***[1] ***94845***[1]	***1877***[1,5] 1509[3] 4864[3]		***E3AY***[1,3]			980	9060
1220											***947BL***[3]			
1222											***947BL***[3,4]			

Hooks that are ***bold italicized*** are barbless.
[1]Daiichi considers this hook to be similar. The brand at the top of the column may or may not concur.
[2]The brand at the top of the column considers the Daiichi hook to the left to be similar. Daiichi may or may not concur.
[3]Similar in all aspects except the bend. See the corresponding characteristic tables.
[4]Similar in all aspects except the finish. See the corresponding characteristic tables.
[5]Similar in all aspects except the point position. See the corresponding characteristic tables.

Daiichi	Dai-Riki	Eagle Claw	Gaelic Supreme	Gamakatsu	Hayabusa	Kamasan	Mustad	Orvis	Owner	Partridge	TMC	Turrall	Varivas	VMC
1250														
1251														
1270	270[1,2,3]	L052[1,2,3]			383	B220[2,4]	80050BR ***80060BR***	1510[1,2,3]			200R[1,3,5] ***200RBL***[3,5]			
1273	270[3,4]	L052[3,4]			383[4]		80050BR[4] ***80060BR***[4]	1510[3,4]			200R[1,3,4,5] ***200RBL***[3,4,5]			
1280	730[1] 300			P10-2L1H[2]	552		94831[1]	1638[1,3]		H1A[1,3]	5212[1]			V7062
1310	305[1,2]						94838[1,3]	1509[1,2,3]		A[3] MM2A[3]	921[1,3]			
1330				S10U[2,4]	375[3]	B440	94842[1,4]							9289[1]
1480	310[2,3]			S13S-M[2,3,4]			94859[1,3]			K1A[1,3]				
1510				C14S[2,3,4]		B160[3]					921			
1530	075[1,2]			S10-2S[2,3,4]			3908[1]	1641			3769[1]			8526[1]
1550	070[2] 305[3]	L057[1,2]			373	B175[2]	3906[1] 3399 3399A 3399N[4] 7948A[5]	1641[1]		G3A[1] CS7MW[3,4] J1A[3]		Sproat ***Barbless Sproat*** Bead Standard Nymph[3]	980[3] 98r[3,4]	8527[1] 8526
1560	060[2]			L10-2H			3906B[1] AC3906B			SH2[3,4]	3761[1]			
1640	310[2]			S12S-1F[2]										
1710	730[1,2]	L063[1,2]		L10-2H[2,3] P10-2L1H[2]		B200[3]	9671[1]	1524[1,2]		D4A[1,3]	5262[1]			9279[1,3]
1720	710[1,2]	L058[1]					9672[1]	1526[1,2]			5263[1]			
1730[6]	700-B[2]			C11-5L2H[2]		B810[3]	81002BR							
1740								594F[1,2]						

Hooks that are ***bold italicized*** are barbless.
[1]Daiichi considers this hook to be similar. The brand at the top of the column may or may not concur.
[2]The brand at the top of the column considers the Daiichi hook to the left to be similar. Daiichi may or may not concur.
[3]Similar in all aspects except the bend. See the corresponding characteristic tables.
[4]Similar in all aspects except the finish. See the corresponding characteristic tables.
[5]Similar in all aspects except the point position. See the corresponding characteristic tables.

Daiichi	Dai-Riki	Eagle Claw	Gaelic Supreme	Gamakatsu	Hayabusa	Kamasan	Mustad	Orvis	Owner	Partridge	TMC	Turrall	Varivas	VMC
1750				S11S-4L2H[2,3]		B940[3,4]	9674[1]	0167[1,2]		D3ST[1,3]	9395[1,3]			9255[1,3]
1770							80150BR	1512[1,2]		K6ST[3]	400T[1,3]			
1850														
1870														
2050				T10-6H[3,4]										
2051				T10-6H[3]										
2052				T10-6H[3,4]										
2055				T10-6H[3,4]										
2059				T10-6H[3]										
2060														
2061														
2062														
2065														
2091										CS10/3[3]				
2101														
2131				T10-6H[3,4]										
2135				T10-6H[3,4]										
2139				T10-6H[3]										
2141														
2151						B220[2,3]								
2161														
2170														
2171														

DAIICHI Cross-Reference

Daiichi	Dai-Riki	Eagle Claw	Gaelic Supreme	Gamakatsu	Hayabusa	Kamasan	Mustad	Orvis	Owner	Partridge	TMC	Turrall	Varivas	VMC
2174														
2220	700[1,2]	L281[1,2]		S11-4L2H[2,3]			79580[1]	8808[1,2] 1511[1,3]		D4A[1] GRS4A[4]	300[1,3]			9283
2271			Salmon Streamer[3,4]				9575[4]			CS17				
2340				Russian River[3]			3665A[1,3]	1511[1,2,3]			300[1,3]			
2370								596F[1,2]						
2421				T10-3H[2]	771[3]	B180[3]	90240[1] 36890[3]	1644[1] 1645[1,3]		N[1,3]	7989[1]	Salmon Single[3]		
2441			Single Low Water[3]	T10-6H[2,3]		B190[3]		1645[1,3]		M[1,3]	7999[1]		974	
2451				S14S-3H[2,3,4] L11S-3H[2,3,4]					5106[3,4]					
2460				L11S-3H[2,3]		B940[4]	3261[1] 3262[4] 80400BLN[3,4] 80400D[3,4]		5320[3,4]					
2461						B940[3]	3262[1,4] 3261[4] 80400BLN[3,4] 80400D[3,4]		5320[3,4]					
2462						B940[4]	3262[1,4] 3261[4] 80400BLN[3] 80400D[3,4]		5320[3]					
2546	930[2]	254SS[1,3]		SS15[2,3,4]			34007[1] 31014[3] 3407[4] 34009 0406[4] 3406B[4] 3366[3,4]	9034		CS51[3,4] CS41[3,4]	811S[1]	Salt Water		9752[3,4]

Hooks that are ***bold italicized*** are barbless.
[1]Daiichi considers this hook to be similar. The brand at the top of the column may or may not concur.
[2]The brand at the top of the column considers the Daiichi hook to the left to be similar. Daiichi may or may not concur.
[3]Similar in all aspects except the bend. See the corresponding characteristic tables.
[4]Similar in all aspects except the finish. See the corresponding characteristic tables.
[5]Similar in all aspects except the point position. See the corresponding characteristic tables.

Daiichi	Dai-Riki	Eagle Claw	Gaelic Supreme	Gamakatsu	Hayabusa	Kamasan	Mustad	Orvis	Owner	Partridge	TMC	Turrall	Varivas	VMC
2550														7299 ***7399*** 7299SS[4]
2552														7299 ***7399*** 7299SS[4]
2553														7299 ***7399*** 7299SS[4]
2556														7299SS 7299[4] ***7399***[4]
2571											105[1]			
2720							37187[1] 80300BR[1]	8810			8089[1,3] 8089NP[3,4]			
2722							37187[1,4]	8810[4]			8089[1,3,4] 8089NP[3]			
3111		L254BK[3] L254N[3,4] L253[3,4] 254N[3,4] 254SS[3,4] 254[3,4] L067[3,4]		SC15[3,4]						MM3ST[3] Z12[3] Z1[3] ***Z1Y***[3] GRS 1Z[3,4] Z2[3,4] Z10[3]				7250NI[3,4] 8410[3,4]
4250														7299 ***7399*** 7299SS[4]
4253														7299 ***7399*** 7299SS[4]
4255														7299[4] ***7399***[4] 7299SS[4]

Daiichi	Dai-Riki	Eagle Claw	Gaelic Supreme	Gamakatsu	Hayabusa	Kamasan	Mustad	Orvis	Owner	Partridge	TMC	Turrall	Varivas	VMC
7131							80525BL[1] 3582F[1]	0528[1,2,3]		Q1[1]		Salmon Double		
X452		250[4]				K93[3,4]		Xpoint						7255[4]
X510							9174[3] 9175[3,4]							

Hooks that are ***bold italicized*** are barbless.
[1]Daiichi considers this hook to be similar. The brand at the top of the column may or may not concur.
[2]The brand at the top of the column considers the Daiichi hook to the left to be similar. Daiichi may or may not concur.
[3]Similar in all aspects except the bend. See the corresponding characteristic tables.
[4]Similar in all aspects except the finish. See the corresponding characteristic tables.
[5]Similar in all aspects except the point position. See the corresponding characteristic tables.

1100 size 16 | 1110 size 12 | 1130 size 10 | 1140 size 18 | 1150, 1155 size 10 | 1170 size 10 | 1180, 1182 size 10 | 1190 size 10 | 1220, 1222 size 14 | 1250, 1251 size 12

Hook	Use	Wire	Shank	Eye	Bend	Finish	Barb	Point	Common Name	Remarks	Size
1100	Dry	1XF	S	TD-O	P	Bron	Mini	Chem	Wide-Gape Dry Fly		16-24
1110	Dry	1XF	S	S-O	P	Bron	Mini	Chem	Wide-Gape Dry Fly, Straight Eye		12-26
1130	Scud/Shrimp	S	1XS	TD	C-R	Bron		Chem	Wide-Gape Scud	Shrimp, grubs, pupae, SJW & nymphs. Formerly J220. Forged.	10-24
1140	Midge	1XF	1XS	TU	C-R	Bron		Chem	Special Wide-Gape	Micro-Caddis. Forged.	18-22
1150	Scud	1XH	S	TU	C-R	Bron		Chem	Heavy Wide-Gape Scud	Shrimp, grubs, pupae, SJW, nymphs. Same as 1155. Forged.	8-18
1155	Scud/Shrimp	1XH	S	TU	C-R	Gold		Chem	Heavy Wide-Gape Scud	Shrimp, grubs, pupae, SJW, nymphs. Same as 1150. Forged.	8-18
1170	Dry	S	S	TD	P	Bron		Chem	Standard Dry Fly	Forged.	8-16
1180	Dry	S	S	TD	P	Bron	Mini	Chem	Standard Dry Fly, Mini-Barb	Same as 1182.	6-24
1182	Dry	S	S	TD	P	Corr-Sil	Mini	Chem	Standard Dry Fly, Mini-Barb	Crystal finish. Same as 1180.	6-24
1190	Dry	S	S	TD	P	Bron	No Barb	Chem	Barbless Dry Fly		8-24
1220	Dry	S	S-BU-M	TD	Spr-R	Bron	Mini	Chem	Darrel Martin's Dry Fly	Same as 1222. Shank is tapered to the eye. Forged.	14-24
1222	Dry	S	S-BU-M	TD	Spr-R	Corr-Sil	Mini	Chem	Darrel Martin's Dry Fly	Crystal finish. Same as 1220. Shank is tapered to the eye. Forged.	14-24
1250	Scud/Shrimp	S	1XL	TU	C-R	Bron	No Barb	Chem	Dennis Brown Glass Bead Fly	Beadheads, pupae. Same as 1251.	12-18
1251	Scud/Shrimp	S	1XL	TU	C-R	Blk	No Barb	Chem	Dennis Brown Glass Bead Fly	Beadheads, pupae. Same as 1250.	12-18

DAIICHI Characteristics

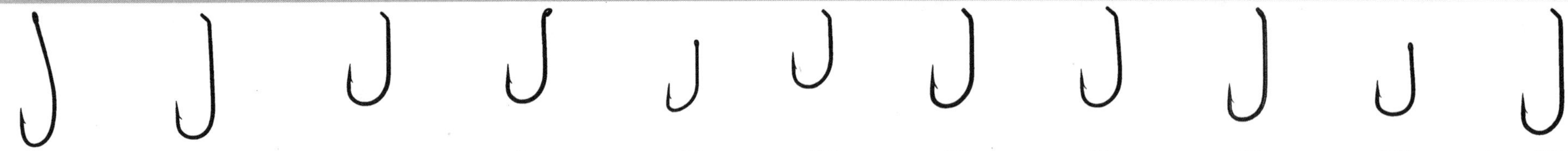

1270, 1273 size 10 | 1280 size 10 | 1310 size 10 | 1330 size 10 | 1480 size 12 | 1510 size 10 | 1530 size 10 | 1550 size 10 | 1560 size 10 | 1640 size 10 | 1710 size 10

Hook	Use	Wire	Shank	Eye	Bend	Finish	Barb	Point	Common Name	Remarks	Size
1270	Nymph	S	3XL-C	S	York	Bron		Chem	Multi-Use Curved	Stoneflies, terrestrials, midges. Same as 1273. Gradual curve along shank. Forged.	4-22
1273	Nymph	S	3XL-C	S	York	Red		Chem	Multi-Use Curved	Stoneflies, nymphs, midges. Same as 1270. Gradual curve along shank. Forged.	8-22
1280	Dry	1XF	2XL	TD	P	Bron	Mini	Chem	2X-Long Dry Fly	Long-bodied dry flies, terrestrials.	6-16
1310	Dry	S	1XS	TD	P	Bron		Chem	Short Shank Dry Fly	Standard dry flies with wide gape, Short bodied dry flies.	8-22
1330	Dry	S	1XS	TU	P	Bron		Chem	Short Shank Dry Fly, Up Eye	Standard dry flies with wide gape, Short bodied dry flies.	8-24
1480	Dry	1XF	2XS	S	Lim	Bron	Mini	Chem	Limerick Dry Fly	Midges, spiders, variants.	12-24
1510	Egg	S	3XS	TD	Spr	Bron		Chem	Glow-Bug	Spiders, ants.	6-18
1530	Wet	2XH	1XS	TD	Spr	Bron		Chem	Heavy Wet Fly	Short-bodied wet flies, nymphs & steelhead flies.	4-16
1550	Wet	S	S	TD	Spr	Bron		Chem	Standard Wet Fly	Nymphs.	2-18
1560	Nymph	1XH	1XL	TD	Spr	Bron		Chem	Traditional Nymph	Wet flies.	6-18
1640	Dry	S	2XS	S	P-R	Bron		Chem	Multi-Use Dry Fly	Caddis, spiders, ants, eggs, tricos, extended-body dry flies, tube flies. Forged.	2-20
1710	Nymph	1XH	2XL	TD	P	Bron		Chem	2X-Long Nymph	Wet flies, Muddlers, terrestrials. Forged.	2-18

Characteristics **DAIICHI**

Hook	Use	Wire	Shank	Eye	Bend	Finish	Barb	Point	Common Name	Remarks	Size
1720	Nymph	1XH	3XL	TD	P	Bron		Chem	3X-Long Nymph hook	Long-bodied nymphs, Muddlers, Woolly Buggers & small streamers. Forged.	4-18
1730	Nymph	1XH	3XL-BD	TD	P	Bron		Chem	Stonefly Nymph	Stonefly nymphs, crabs. Forged.	4-14
1740	Nymph	2XH	2XL	TU	P	Bron		Chem	Up-Eye Nymph		6-16
1750	Streamer	1XH	4XL	S	P	Bron		Chem	Straight-Eye Streamer	Muddlers, Zonkers, bucktails. Forged.	4-14
1770	Nymph	1XF	3XL-BU	S	Spr	Bron		Chem	Swimming Nymph	Leeches & SJW. Shank curved up. Forged.	6-16
1850	Streamer	1XH	4XL	F	P	Bron		Chem	Flat-Eye Streamer	Free-swimming streamer patterns. Forged.	6-12
1870	Nymph	2XH	3XL-BD	TU	York	Bron		Chem	Larva	Swimming nymphs and larval patterns. Shank curved down.	4-14

2050, 2051, 2052, 2055, 2059, 2060, 2061, 2062, 2065
size 5

2091
size $1^1/_2$

Hook	Use	Wire	Shank	Eye	Bend	Finish	Barb	Point	Common Name	Remarks	Size
2050	Steelhead/ Salmon	S	S-C	TU-L	Lim-M	Bron		Chem	Alec Jackson Spey Fly	Same as 2051, 2052, 2055 & 2059. Odd-numbered sizes. Forged.	3/0-7
2051	Steelhead/ Salmon	S	S-C	TU-L	Lim-M	Blk		Chem	Alec Jackson Spey Fly	Same as 2050, 2052, 2055 & 2059. Odd-numbered sizes. Forged.	3/0-7
2052	Steelhead/ Salmon	S	S-C	TU-L	Lim-M	Nick		Chem	Alec Jackson Spey Fly	Same as 2050, 2051, 2055 & 2059. Odd-numbered sizes. Forged.	3/0-7
2055	Steelhead/ Salmon	S	S-C	TU-L	Lim-M	Gold		Chem	Alec Jackson Spey Fly	Same as 2050, 2051, 2052 & 2059. Odd-numbered sizes. Forged.	3/0-7
2059	Steelhead/ Salmon	S	S-C	TU-L	Lim-M	Blue		Chem	Alec Jackson Spey Fly	Same as 2050, 2051, 2052 & 2055. Odd-numbered sizes. Forged.	3/0-7
2060	Steelhead/ Salmon	1XH	S-C	TU-L	Lim-M	Bron		Chem	Alec Jackson Spey Fly	Same as 2061, 2062, 2065. Forged.	3
2061	Steelhead/ Salmon	1XH	S-C	TU-L	Lim-M	Blk		Chem	Alec Jackson Spey Fly	Same as 2060, 2062, 2065. Forged.	3
2062	Steelhead/ Salmon	1XH	S-C	TU-L	Lim-M	Nick		Chem	Alec Jackson Spey Fly	Same as 2060, 2061, 2065. Forged.	3
2065	Steelhead/ Salmon	1XH	S-C	TU-L	Lim-M	Gold		Chem	Alec Jackson Spey Fly	Same as 2060, 2061, 2062. Forged.	3
2091	Display	S	1XL-C	No	Lim-M	Blk		Chem	Alec Jackson Blind-Eye Fly Hook	Same as 2051 but the tapered loop eye is straightened. Hence, 1XL shank. Forged.	$1^1/_2$

Hook	Use	Wire	Shank	Eye	Bend	Finish	Barb	Point	Common Name	Remarks	Size
2101	Streamer	4XH	S	S-L	Lim-M	Blk		Chem	Alaska Salmon	Streamers. Formerly J101.	6
2131	Steelhead/ Salmon	S	S-C	TU-L	Lim-M-R	Blk		Chem	Bob Veverka's "Classic Salmon"	Slight curve & offset point. Same as 2135 & 2139.	2-8
2135	Steelhead/ Salmon	S	S-C	TU-L	Lim-M-R	Gold		Chem	Bob Veverka's "Classic Salmon"	Slight curve & offset point. Same as 2131 & 2139.	2-8
2139	Steelhead/ Salmon	S	S-C	TU-L	Lim-M-R	Blue		Chem	Bob Veverka's "Classic Salmon"	Slight curve & offset point. Same as 2131 & 2135.	2-8
2141	Steelhead/ Salmon	1XH	S	S-L	Lim	Blk		Chem	Straight-Eye Salmon	Streamers, Muddlers, Woolly Buggers & bonefish patterns. Forged. Formerly J141.	1-6
2151	Steelhead/ Salmon	S	S-C	S-L	Lim-M	Blk		Chem	Curved Shank Salmon	Nymphs. Forged.	1-10
2161	Steelhead/ Salmon	S	S-C	TU-L	Lim-M	Blk		Chem	Curved Shank Salmon - Up-Eye	Nymphs. Forged.	1-6
2170	Egg	S	1XS-BU	S	P-R	Bron		Chem	Bent-Shank	Emergers, SJW. Same as 2171 & 2174. Forged.	2-12
2171	Egg	S	1XS-BU	S	P-R	Blk		Chem	Bent-Shank	Emergers, SJW. Same as 2170 & 2174. Forged.	2-12
2174	Egg	S	1XS-BU	S	P-R	Grn		Chem	Bent-Shank	Emergers, SJW. Same as 2170 & 2171. Forged.	2-12
2220	Streamer	1XH	4XL	TD	P	Bron		Chem	4X-Long Streamer	Muddlers, Woolly Buggers, Zonkers, bucktails. Forged.	1-14
2271	Steelhead/ Salmon	1XH	6XL	TD-L	Lim-M	Blk		Chem	Dee/Streamer	Spey flies, salmon streamers. Forged. Formerly J171.	1, 2

DAIICHI Characteristics

Hook	Use	Wire	Shank	Eye	Bend	Finish	Barb	Point	Common Name	Remarks	Size
2340	Streamer	1XH	6XL	TD	Lim	Bron		Chem	6X-Long Streamer		4-12
2370	Streamer	3XH	7XL	S-L	Lim-M	Bron		Chem	7X-Long Streamer	Designed by Dick Talleur.	2-10
2421	Steelhead/ Salmon	S	S	TU-L	Spr-M	Blk		Chem	Low Water Salmon	Low water patterns and steelhead/salmon dry flies. Forged.	2-12
2441	Steelhead/ Salmon	1XH	S	TU-L	Spr-M	Blk		Chem	Traditional Salmon	Classic salmon wet flies and hairwing steelhead/salmon. Forged.	2/0-8
2451	Steelhead/ Salmon	S	1XS	S	Osh	Blk		Chem	Short Shank Salmon	Salmon trailer, bass, tube flies. Forged.	4/0-8
2460	Bass/Pike	S	3XL	S	Aber	Bron		Chem	Multi-Use Aberdeen	Long-bodied dry flies, beadheads. Same as 2461 & 2462.	6/0-18
2461	Bass/Pike	S	3XL	S	Aber	Blk		Chem	Multi-Use Aberdeen	Long-bodied dry flies, beadheads. Same as 2460 & 2462.	6/0-18
2462	Bass/Pike	S	3XL	S	Aber	Nick		Chem	Multi-Use Aberdeen	Long-bodied dry flies, beadheads. Same as 2460 & 2461.	6/0-18
2546	Saltwater	S	S	S	Osh	SS			Salt Water	Bonefish, tarpon, tube flies. Forged.	8/0-6
2550	Salmon	S	1XS	TU	Beak-R	Bron		Chem	Octopus Salmon	Same as 2552, 2553 & 2556.	6-2
2552	Salmon	S	1XS	TU	Beak-R	Nick		Chem	Octopus Salmon	Same as 2550, 2553 & 2556.	6-2
2553	Salmon	S	1XS	TU	Beak-R	Red		Chem	Octopus Salmon	Same as 2550, 2552 & 2556.	6-2
2556	Salmon	S	1XS	TU	Beak-R	SS		Chem	Octopus Salmon	Same as 2550, 2552 & 2553.	6/0-6

Hook	Use	Wire	Shank	Eye	Bend	Finish	Barb	Point	Common Name	Remarks	Size
2571	Steelhead	2XH	3XS	TD-M	Spr-R	Blk		Chem	Boss Steelhead	Egg patterns. Eye partially TD. Forged.	2-8
2720	Bass	1XF	S	S	WG	Bron		Chem	Wide Gape "Stinger"	Deer hair bass bugs, divers, frogs, mice. Same as 2722.	5/0-2
2722	Bass	1XF	S	S	WG	Nick		Chem	Wide Gape "Stinger"	Deer hair bass bugs, divers, frogs, mice. Same as 2720.	5/0-2
3111	Salmon	S	S	S	Beak-R	Blk		Chem	Octopus Salmon		10/0-1
4250	Egg	S	1XS	TU	Beak-R	Bron		Chem	Salmon Egg	Same as 4253 & 4255. Forged.	4-14
4253	Egg	S	1XS	TU	Beak-R	Red		Chem	Salmon Egg	Same as 4250 & 4255. Forged.	4-14
4255	Egg	S	1XS	TU	Beak-R	Gold		Chem	Salmon Egg	Same as 4250 & 4253. Forged.	4-14
7131	Double	S	S	TU-L	Lim-Dbl	Blk		Chem	Double Salmon	Forged.	4-12
X452	Saltwater	S	1XL	S	Osh	Sil		Chem	X-Point Salt Water	Salmon trailer, bass, tube flies. Forged.	4/0-6
X510	Steelhead/ Salmon	3XH	3XS	S	WG	Bron		Chem	X-Point Steelhead	Eggs, short steelhead/ salmon flies. Forged.	4-8

Dai-Riki	Daiichi	Eagle Claw	Gaelic Supreme	Gamakatsu	Hayabusa	Heritage	Kamasan	Mustad	Orvis	Owner	Partridge	TMC	Turrall	Varivas	VMC
060	1560[1]						B170[3]	3906B[1,3] AC3906B[1] 7957B			SH2[3,4]	3761[1] ***3761BL***	Extra-long Nymph Bead Extra-long Nymph[3]		8527[3]
070	1550[1]	L060						3906[1]				3769[1]	Sproat ***Barbless Sproat*** Bead Standard Nymph[3]		
075	1530[1,2]	L057						3906[1] 3908	1641[2] 167T		G3A	3769[1,4] 9300[3]		977[3,4]	
135	1150[1] 1130[1]		Grub[3,4]	C12	387[3]	Grub[3,4]	B110[3]	80200BR[1,3] 80250BR[3]	1639[2]		K4A	2457[1,3]			
270	1270[1,2,3] 1273[3,4]				383[3]			80050BR[1,3]	1510[2,3]			200R[1,3,5] ***200RBL***[3,5]			
300	1170[1,2]	L059						94833[1] 7957B 94840 AC94840			***CS28GRS***[3,4]	100[1,3] 5210[3]	Extra-long Nymph Bead Extra-Long[3]		8527 9280[3]
305	1310[1,2] 1170 1180 1182[4] ***1190*** 1550[3]	L059		P-10 ***R10-BN***[3,4] S10[4] ***S10-B***[4]	***378*** 373[3]		***B402*** K40	94840[1] AC94840[1] 3399[3] 3399A[3] 3399N[3,4]	1509[3] 4864[3]		L2A[3]	5210[1] 100[1,3]		980	9060
310	1640[1,3] 1480[1,3]			S12S-1F[3] S13S-M[3,4]	377 ***379***			94859[1]	4641[2,3]			101[1,3]			
700	2220[1,2]	L281						79580[1]	8808		D4A GRS4A[4]				9283
700-B[6]	1730[1,2]			C11-5L2H			B810	81002BR							
710	1720[1]	L058						9672[1] AC9672[1]	1526		SH3[3,4]	5263[1]	Extra-long Streamer ***Barbless Extra-long*** Bead Streamer[3]		9283
730	1710[1,2] 1280[2]	L063		P10-2L1H	552			9671[1] AC9671[1]	1524[2]		H1A[3]	5262[1] 5212			V7062 9279[3]

Dai-Riki	Daiichi	Eagle Claw	Gaelic Supreme	Gamakatsu	Hayabusa	Heritage	Kamasan	Mustad	Orvis	Owner	Partridge	TMC	Turrall	Varivas	VMC
930	2546[1]	L054SS 254SS[3] L067[3,4,5] L7254[4]						34007[1]	9034	5170[4]		811S[1] 800S[5]	Salt Water		9752[3,4]

Hooks that are ***bold italicized*** are barbless.
[1] Dai-Riki considers this hook to be similar. The brand at the top of the column may or may not concur.
[2] The brand at the top of the column considers the Dai-Riki hook to the left to be similar. Dai-Riki may or may not concur.
[3] Similar in all aspects except the bend. See the corresponding characteristic tables.
[4] Similar in all aspects except the finish. See the corresponding characteristic tables.
[5] Similar in all aspects except the point position. See the corresponding characteristic tables.
[6] Weight and length varies. Eye, bend, finish and shape are similar unless otherwise noted.

DAI-RIKI Characteristics

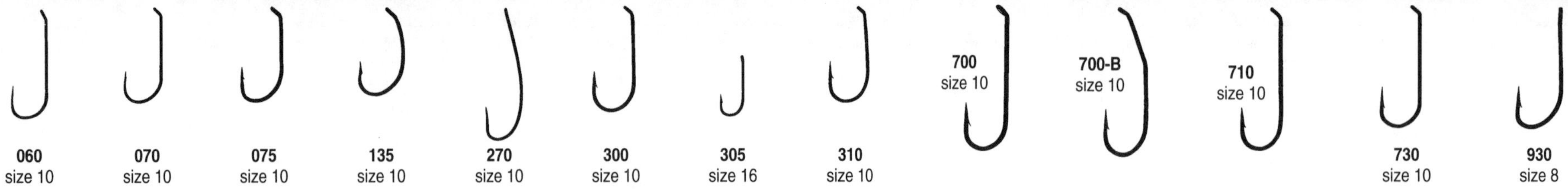

Hook	Use	Wire	Shank	Eye	Bend	Finish	Barb	Point	Common Name	Remarks	Size
060	Nymph	1XH	1XL	TD	Spr	Bron	Mini	Chem	Nymph	Wet flies.	6-22
070	Wet	1XF	S	TD	Spr	Bron	Mini	Chem	Wet Fly	Nymphs, shrimp, emergers.	6-20
075	Nymph	2XH	S	TD	Spr	Bron	Mini	Chem	Wet Fly	Wet flies, steelhead & shrimp.	8-20
135	Scud/Shrimp	1XH	1XS-C	TD	C-WG-R	Bron	Mini	Chem	Scud	Caddis larva, pupae.	6-22
270	Nymph	S	3XL-C	S	Lim-M	Bron	Mini	Chem	Natural Bend	Dry flies.	4-24
300	Dry	S	1XL	TD	P	Bron	Mini	Chem	Dry Fly or Light Nymph	Nymphs.	8-18
305	Dry	S	S	TD	P	Bron	Mini	Chem	Dry Fly	Emergers, nymphs.	8-22
310	Dry	S	S	S	P	Bron	Mini	Chem	Dry Fly	Spinners.	16-22
700	Streamer	1XH	4XL	TD	P	Bron	Mini	Chem	Streamer	Large nymphs.	1-16
700-B	Nymph	1XH	4XL-BD	TD	P	Bron	Mini	Chem	Nymph	Caddis larva.	2-16
710	Streamer	1XH	3XL	TD	P	Bron	Mini	Chem	Streamer	Nymphs, Woolly Buggers.	2-16
730	Nymph	1XF	2XL	TD	P	Bron	Mini	Chem	Nymph	Hoppers.	2-18
930	Saltwater	1XH	S	S	Osh	SS	Mini	Chem	Salt Water	Bass.	3/0-1/0, 2-8

Eagle Claw	Daiichi	Dai-Riki	Gaelic Supreme	Gamakatsu	Hayabusa	Kamasan	Mustad	Orvis	Owner	Partridge	TMC	Turrall	Varivas	VMC
066														
066N														
066SS														
072														
178							3408B[3,4]							
250	X452[4]					K93[3,4]	3407[1]	XPoint[3,4]		CS52[1,3,4]				7255[4]
250H										S4[3,4] GRS S4[3,4]				
254	3111[3,4]			SC15[4]						MM3ST[3,4] Z12[3,4] Z1[3,4] ***Z1Y***[3,4] GRS 1Z[3,4] Z2[3,4] Z10[3,4]				7250NI[3,4] 8410[4]
254N	3111[3,4]			SC15[4]						MM3ST[3,4] Z12[3,4] Z1[3,4] ***Z1Y***[3,4] GRS 1Z[3,4] Z2[3,4] Z10[3,4]				7250NI[3] 8410[4]
254SS	2546[2] 3111[3,4]	930[4]		SC15[3,4] SS15			34007 31014 3407[3,4]	9034		CS51 MM3ST[3,4] Z12[3,4] Z1[3,4] ***Z1Y***[3,4] GRS 1Z[3,4] Z2[3,4] Z10[3,4]		Salt Water[3]		9752[3,4] 7250NI[3,4] 8410[3,4]
L052	1270[1,2,3] 1273[3,4]				383[3]		80050BR[1,3]	1510[3]		K12ST[1,3]	200R[1,3,5] ***200RBL***[3,5]			
L054SS		930					77660SS[1]		5170[3,4]		811S[1] 800S[5]			

Hooks that are ***bold italicized*** are barbless.
[1]Eagle Claw considers this hook to be similar. The brand at the top of the column may or may not concur.
[2]The brand at the top of the column considers the Eagle Claw hook to the left to be similar. Eagle Claw may or may not concur.
[3]Similar in all aspects except the bend. See the corresponding characteristic tables.
[4]Similar in all aspects except the finish. See the corresponding characteristic tables.
[5]Similar in all aspects except the point position. See the corresponding characteristic tables.

Eagle Claw	Daiichi	Dai-Riki	Gaelic Supreme	Gamakatsu	Hayabusa	Kamasan	Mustad	Orvis	Owner	Partridge	TMC	Turrall	Varivas	VMC
L055	1130[1,2,3]						80250BR[1,3]			K4A[1,3]	2487[1,3]			
L056				C12			80200BR[1,3]				2457[1]	Grub		9902[3]
L057	1550[1,2]	070		L10-2H	373		3906[1] 3399 3399A 3399N[4] 7948A			G3A[1] CS7MW[3,4] J1A[3]	3761[1]	Sproat ***Barbless Sproat*** Bead Standard Nymph[3]	98r[3,4]	8527 8526
L058	1720[2]	710					9672[1]			D4A	5363[1]			
L059	1180[1] 1170[2] 1100			P-10[1] S10-3F[3]	572[3]	B400[3]	94840[1] AC94840 ***94845*** 80000BR ***80010BR***	1523		L2A[1,3] L3A[3] ***L3AY***[3] GRS3A[3,4]	9300[1,3] 5210 100[3] ***100BL***[3] 146Z[4] 102Y[3,4] ***103BL***[3,4] ***900BL***[3,4,5]			V7061 9280[3]
L060	1100[3]	070		S10-3F[3,4]	572	B400[3]	3399[1] 94840[3] AC94840[3] ***94845***[3] 80000BR[3] ***80010BR***[3]	1523[3]		L2A[1,3] L3A[3] ***L3AY***[3] GRS3A[3,4]	5212[1,3] 5210[3] 100[3] ***100BL***[3] 146Z[3,4] 102Y[3,4] ***103BL***[3,4] ***900BL***[3,4,5]			V7061[3] 9280[3]
L061B	***1190***[1,2] 1100			S10-3F[3]	572[3]	B400[3]	***94845***[1] 94840 AC94840 80000BR ***80010BR***	1523		E1AY[1] L2A[3] L3A[3] ***L3AY***[3] GRS3A[3,4]	100[1,3] ***100BL***[3] 5210 146Z[4] 102Y[3,4] ***103BL***[3,4] ***900BL***[3,4,5]			V7061 9280[3]
L063	1710[1,2]			P10-2L1H[1]		B830	9671[1] AC9671	1524 1638[3]		H1A[1,3]	5262[1]			9279[3]
L064B											***905BL***			

Hooks that are ***bold italicized*** are barbless.
[1]Eagle Claw considers this hook to be similar. The brand at the top of the column may or may not concur.
[2]The brand at the top of the column considers the Eagle Claw hook to the left to be similar. Eagle Claw may or may not concur.
[3]Similar in all aspects except the bend. See the corresponding characteristic tables.
[4]Similar in all aspects except the finish. See the corresponding characteristic tables.
[5]Similar in all aspects except the point position. See the corresponding characteristic tables.

Eagle Claw	Daiichi	Dai-Riki	Gaelic Supreme	Gamakatsu	Hayabusa	Kamasan	Mustad	Orvis	Owner	Partridge	TMC	Turrall	Varivas	VMC
L067	2546[3,4] 3111[3,4]			SC15[3,4] SS15			34007[3,4,5] 31014[3,4,5] 3407[3,4,5]	9034[3,4]		CS51[3,4] MM3ST[3,4] Z12[3,4] Z1[3,4] ***Z1Y***[3,4] GRS 1Z[3,4] Z2[3,4] Z10[3,4]	800S[1,3,4,5]	Salt Water[3,4]		9752[3,4] 7250NI[3,4] 8410[3,4]
L072BK														
L1197B						B175	7970[3]	167T			700[3,4,5] 9300[3]			
L1197G						B175[4]	7970[3,4]	167T[4]			700[3,4,5] 9300[3,4]			
L1197N						B175[4]	7970[3,4]	167T[4]			700[3,4,5] 9300[3,4]			
L144	1150[1]		Caddis[3,4]	C12U[1]	***388***[5]		37160[1,3] 37161[3,4] 37162[3,4]				***205BL***[1,3,5]			
L200							33900[1,3] 33903[3] 34015D[3] 90233S[3,4]				511S[3,4]			
L253	3111[3,4]			SC15[3,4]						MM3ST[3,4] Z12[3,4] Z1[3,4] ***Z1Y***[3,4] GRS 1Z[3,4] Z2[3,4] Z10[3,4]				7250NI[3,4] 8410[4]
L254BK	3111[3]			SC15[3,4]						MM3ST[3] Z12[3] Z1[3] ***Z1Y***[3] GRS 1Z[3,4] Z2[3,4] Z10[3]				7250NI[3,4] 8410[4]

Eagle Claw	Daiichi	Dai-Riki	Gaelic Supreme	Gamakatsu	Hayabusa	Kamasan	Mustad	Orvis	Owner	Partridge	TMC	Turrall	Varivas	VMC
L254N	3111[3,4]			SC15[3,4]						MM3ST[3,4] Z12[3,4] Z1[3,4] ***Z1Y***[3,4] GRS 1Z[3,4] Z2[3,4] Z10[3,4]				7250NI[3] 8410[4]
L256							9174[1]							
L281	2220[1,2]	700		S11-4L2H[1,3]			79580[1]			D4A[1] GRS4A[4]	5263[1]			
L7254		930[4]							5170[3,4]		811S[4] 800S[4,5]			
NT1197						B175[4]	7970[3,4]	167T[4]			700[3,4,5] 9300[3,4]			
NT156							9174[4]							

Hooks that are ***bold italicized*** are barbless.
[1]Eagle Claw considers this hook to be similar. The brand at the top of the column may or may not concur.
[2]The brand at the top of the column considers the Eagle Claw hook to the left to be similar. Eagle Claw may or may not concur.
[3]Similar in all aspects except the bend. See the corresponding characteristic tables.
[4]Similar in all aspects except the finish. See the corresponding characteristic tables.
[5]Similar in all aspects except the point position. See the corresponding characteristic tables.

066, 066N, 066SS, 072, L072BK
size 10

178
size 10

250
size 6

250H
size 9/0

254, 254N, 254SS, L254BK, L7254, L254N
size 4

L052
size 10

Hook	Use	Wire	Shank	Eye	Bend	Finish	Barb	Point	Common Name	Remarks	Size
066	Saltwater	S	2XL	S	Lim-M-R	Corr-Sil				Bass. Same as 066N, 066SS, 072, L072BK. Forged.	7/0-6
066N	Saltwater	S	2XL	S	Lim-M-R	Nick				Bass. Same as 066, 066SS, 072, L072BK. Forged.	5/0-8
066SS	Saltwater	S	2XL	S	Lim-M-R	SS				Bass. Same as 066, 066N, 072, L072BK. Forged.	6/0-4
072	Bass	S	2XL	S	Lim-M-R	Bron				Saltwater. Same as 066, 066N, 066SS, L072BK. Forged.	4/0-10
178	Saltwater	S	S	TD	Spr-M	Gold				Bass. Forged.	2-10
250	Saltwater	S	1XL	S	Osh	Corr-Sil		Chem		SeaGuard finish. Forged.	12/0-6
250H	Saltwater	2XH	1XL	S	Osh	Corr-Sil		Chem		SeaGuard finish. Forged.	10/0-1/0
254	Saltwater	S	S	S	Osh	Corr-Sil				Same as 254N, L254N, L254BK. Tube flies. Forged.	14/0-1/0
254N	Saltwater	S	S	S	Osh	Nick				Same as 254, L254N, L254BK. Tube flies. Forged.	14/0-2
254SS	Saltwater	S	S	S	Osh-R	SS				Bass. Forged.	9/0-2
L052	Nymph	1XF	S-C	S	Spr	Bron		Chem		Nickel Teflon finish available. Forged.	6-20

Characteristics

L054SS size 1/0

L055 size 10

L056 size 10

L057 size 10

L058 size 10

L059, L061B size 10

L060 size 10

L063 size 10

Hook	Use	Wire	Shank	Eye	Bend	Finish	Barb	Point	Common Name	Remarks	Size
L054SS	Saltwater	1XH	S	S	Osh	SS		Chem			4/0-1/0
L055	Scud/Shrimp	1XF	1XS	TD	C	Bron		Chem		Nickel Teflon finish available.	10-20
L056	Egg	S	S-C	TD	C	Bron		Chem		Nickel Teflon finish available.	2-14
L057	Nymph	S	S	TD	Spr	Bron		Chem		Nickel Teflon finish available. Forged.	4-20
L058	Streamer	1XH	3XL	TD	P	Bron		Chem		Forged.	4-14
L059	Dry	1XF	S	TD	P	Bron		Chem		Same as L061B. Nickel Teflon finish available. Forged.	4-24
L060	Dry	1XF	S	TD	Spr	Bron		Chem		Wet flies. Nickel Teflon finish available. Forged.	8-18
L061B	Dry	1XF	S	TD	P	Bron	No Barb	Chem		Same as L059. Forged.	4-18
L063	Wet	S	2XL	TD	P	Bron		Chem		Nickel Teflon finish available. Forged.	4-20

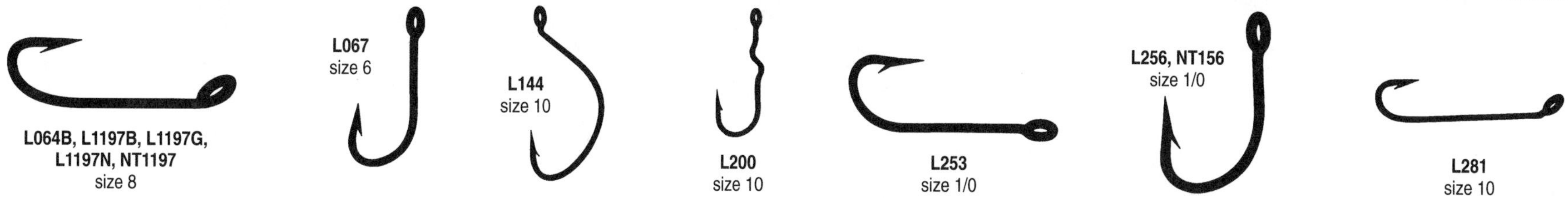

Hook	Use	Wire	Shank	Eye	Bend	Finish	Barb	Point	Common Name	Remarks	Size
L064B	Steelhead/ Salmon	S	S	TD	Spr-M	Bron	No Barb	Chem		Same as L1197N, L1197B & L1197G.	1-8
L067	Saltwater	S	S	S	Spr	Corr-Sil		Spec-C	Billy Pate Saltwater	SeaGuard finish. Straight point. Forged.	5/0-6
L072BK	Saltwater	S	2XL	S	Lim-M-R	Blk-Nick		Chem		Saltwater. Same as 066, 066N, 066SS, 072. Forged.	4/0-6
L1197B	Steelhead/ Salmon	1XH	S	TD	Spr	Bron		Chem		Same as L1197N & L1197G, NT1197.	1-8
L1197G	Steelhead/ Salmon	1XH	S	TD	Spr	Gold		Chem		Same as L1197B & L1197N, NT 1197.	1-8
L1197N	Steelhead/ Salmon	1XH	S	TD	Spr	Nick		Chem		Same as L1197B & L1197G, NT1197.	1-8
L144	Nymph	S	S-C	TU	C-WG	Bron		Chem	Kahle/Nymph		4-16
L200	Bass	S	S-PK	S	P	Bron		Chem	Popper Bug	Poppers.	1/0-12
L253	Saltwater	S	S	S	Osh	Nick		Chem		Tube flies. Forged.	5/0-1/0
L254BK	Saltwater	S	S	S	Osh	Blk		Chem		Blk nick finish. Same as 254, 254N, L254N. Tube flies. Forged.	7/0-1
L254N	Saltwater	S	S	S	Osh	Nick		Chem		Same as 254, 254N, L254BK.Tube flies. Forged.	7/0-1
L256	Saltwater	1XH	2XS	S	Osh-R	Bron		Chem		Same as NT156. Forged.	4/0-4
L281	Streamer	1XH	4XL	TD	P	Bron		Chem		Nickel Teflon finish available. Forged.	2-10
L7254	Saltwater	1XH	S	S	Osh	(Rmks)-Nick		Chem		Teflon coated. Tube flies. Forged.	7/0-4
NT1197	Steelhead/ Salmon	1XH	S	TD	Spr	Nick		Chem		Teflon coated. Same as L1197B & L1197G, L1197N.	1-8
NT156	Saltwater	1XH	2XS	S	Osh-R	Nick		Chem		Teflon coated. Same as L256. Forged.	4/0-1/0

Gaelic Supreme	Daiichi	Dai-Riki	Eagle Claw	Gamakatsu	Hayabusa	Heritage	Kamasan	Mustad	Orvis	Partridge	TMC	Turrall	VMC
Caddis			L144[3,4]	C12U[3,4]	*388*[5]			37160[4] 37161[4] 37162[4]			***205BL***[3,4,5]		
Double Low Water							B280[3]			Q1[3] GRSQ2[3,4]		Salmon Double[3] Sproat Double[3,4] Salmon Double Low-water[3]	9410[3,4]
Double Wet							B280[3]			P[3] Q1[3] GRSQ2[3,4]		Salmon Double[3] Sproat Double[3,4] Salmon Double Low-water[3]	9410[3,4]
Dry fly				S10U[4]				94842[4]		L3B[3,4]			9281[3,4]
F-F Blind Fly												Salmon Blind Eye	
F-F Dee Style Blind Eye						Harrison-Bartleet							
F-F Heavy Dee													
F-F Heavy Spey													
F-F Light Blind Eye						Pryce Tannant				CS6		Salmon Blind Eye	
F-F Light Dee													
F-F Light Spey													
F-F Light Wire								90240[3]				Salmon Extra-fine Dry Fly[3]	
F-F Low Water						Single Salmon Low Water		80500BL[3]				Salmon Low-water[3]	
F-F Spey Style Blind Eye													
F-F Wet Fly										M[3]			
Grub Hook		135[3,4]			387[3,4]	Grub[4]		80200BR[3,4]			2457[3,4]		
Long Shank Dry Fly													

Gaelic Supreme	Daiichi	Dai-Riki	Eagle Claw	Gamakatsu	Hayabusa	Heritage	Kamasan	Mustad	Orvis	Partridge	TMC	Turrall	VMC
Mike Martinek Rangeley Streamer 6XL (MMRS-6XL)						Allcock Streamer		33957	1511[3]	D1A[3]			
Mike Martinek Rangeley Streamer 8XL (MMRS-8XL)													
Mike Martinek Rangeley Streamer 10XL (MMRS-10XL)										CS15[3]			
Nymph Hook						Nymph Hook							
Single Dry fly									1644[3]		7989[3]	Salmon Extra-fine[3]	
Single Low Water	2441[3]						B190[3]		1645[3]	M[3,4]	7999[3]		
Single Wet	2441[3]					Single Salmon		36890[3]		M[3,4]	7999[3]		
Steelhead Egg Hook													
Trout Streamer													
Salmon Streamer	2271[3,4]							9575[3]		CS17[3,4]			
Treble Low Water							B380[3]			X2B[3,5] CS12[3] GRSX2B[3,4,5]		Salmon Treble[3,5] Salmon Treble Low-water[3,5]	
Treble Wet							B380[3]			X2B[3,5] GRSX2S[3,4,5] CS12[3]		Salmon Treble[3,5] Salmon Treble Low-water[3,5]	
Wet Fly Hook				S10U[3,4]									
Wide Gape Midge													

Hooks that are ***bold italicized*** are barbless.
[1]Gaelic Supreme considers this hook to be similar. The brand at the top of the column may or may not concur.
[2]The brand at the top of the column considers the Gaelic Supreme hook to the left to be similar. Gaelic Supreme may or may not concur.
[3]Similar in all aspects except the bend. See the corresponding characteristic tables.
[4]Similar in all aspects except the finish. See the corresponding characteristic tables.
[5]Similar in all aspects except the point position. See the corresponding characteristic tables.

Hook	Use	Wire	Shank	Eye	Bend	Finish	Barb	Point	Common Name	Remarks	Size
Caddis	Scud/Shrimp	S	S-C	TU	C	Olive	Mini	Chem	Gaelic Supreme Trout Caddis Hooks	Forged.	8-16
Double Low Water	Double	1XH	2XL	TU-L	N.I.-Dbl	Blk		Chem	Gaelic Supreme Salmon Double Low Water	Forged.	2/0-12
Double Wet	Double	2XH	S	TU-L	N.I.-Dbl	Blk		Chem	Gaelic Supreme Salmon Double Wet	Forged.	2/0-12
Dry Fly	Dry	4XF	S	TU-M	P	M-Nick	Mini	Chem	Gaelic Supreme Trout Dry Fly	Smaller sizes have 2X fine wire. Forged.	10-20
F-F Blind Eye	Display	S	S	No	DLim	Blk			Gaelic Supreme Francis-Francis Salmon Blind Eye		6/0-1/0
F-F Dee Style Blind Eye	Display	S	5XL	No	DLim	Blk			Gaelic Supreme Francis-Francis Salmon Dee Style Blind Eye		3/0-1/0
F-F Heavy Dee	Steelhead/ Salmon	2XH	5XL	TU-L	DLim	Blk		Chem	Gaelic Supreme Francis-Francis Salmon Heavy Dee	Forged.	3/0-1/0, 1½-7
F-F Heavy Spey	Steelhead/ Salmon	2XH	3.5XL	TU-L	DLim	Blk		Chem	Gaelic Supreme Francis-Francis Salmon Heavy Spey	Spey flies. Forged.	3/0-1/0, 1½-7
F-F Light Blind Eye	Display	2XF	1XL	No	DLim	Blk			Gaelic Supreme Francis-Francis Salmon Light Blind Eye		6/0-1/0
F-F Light Dee	Steelhead/ Salmon	S	5XL	TU-L	DLim	Blk		Chem	Gaelic Supreme Francis-Francis Salmon Light Dee	Forged.	3/0-1/0, 1½-7
F-F Light Spey	Steelhead/ Salmon	S	3.5XL	TU-L	DLim	Blk		Chem	Gaelic Supreme Francis-Francis Salmon Light Spey	Spey flies. Forged.	3/0-1/0, 1½-7
F-F Light Wire	Steelhead/ Salmon	2XF	1XL	TU-L	DLim	Blk		Chem	Gaelic Supreme Francis-Francis Salmon Light Wire	Forged.	3/0-1/0, 1½-7
F-F Low Water	Steelhead/ Salmon	S	2XL	TU-L	DLim	Blk		Chem	Gaelic Supreme Francis-Francis Salmon Low Water	Forged.	3/0-1/0, 1½-7

Hook	Use	Wire	Shank	Eye	Bend	Finish	Barb	Point	Common Name	Remarks	Size
F-F Spey Style Blind Eye	Display	S	3.5XL	No	DLim	Blk			Gaelic Supreme Francis-Francis Salmon Spey Style Blind Eye		3/0-1/0
F-F Wet Fly	Steelhead/ Salmon	2XH	S	TU-L	DLim	Blk		Chem	Gaelic Supreme Francis-Francis Salmon Wet Fly	Forged.	3/0-1/0, 1$^{1}/_{2}$-7
Grub Hook	Scud/Shrimp	1XH	S-C	TD	C	Olive	Mini	Chem	Gaelic Supreme Trout Grub Hook	Grubs. Forged.	8-16
Long Shank Dry Fly	Dry	4XF	2XL	TU-M	P	M-Nick	Mini	Chem	Gaelic Supreme Trout Long Shank Dry Fly	Eye partially TD. Matte finish. Forged.	8-12
Mike Martinek Rangley Streamer 6XL (MMRS-6XL)	Streamer	S	6XL	TD-M	Spr-M	Bron			Mike Martinek Rangley Streamer 6XL	Modeled after a Carrie Stevens hook. Martinek modified the eye to a 30° turn.	1, 2, 4
Mike Martinek Rangley Streamer 8XL (MMRS-8XL)	Streamer	2XH	8XL	TD-M	Spr-M	Bron			Mike Martinek Rangley Streamer 8XL	Modeled after a Carrie Stevens hook. Martinek modified the eye to a 30° turn.	1, 2, 4
Mike Martinek Rangley Streamer 10XL (MMRS-10XL)	Streamer	2XH	10XL	TD-M	Spr-M	Bron			Mike Martinek Rangley Streamer 10XL	Modeled after a Carrie Stevens hook. Martinek modified the eye to a 30° turn. Wire size is 2XH except in size 4, which is standard size.	1, 2, 4
Nymph Hook	Nymph	S	3XL	TU-M	P	Olive	Mini	Chem	Gaelic Supreme Trout Nymph Hook	Eye partially TD. Forged.	8-16
Salmon Streamer	Steelhead/ Salmon	1XH	7XL	TD-L	Spr	Bron		Chem	Gaelic Supreme Salmon Streamer Hook	Streamers. Forged.	1, 2, 3
Single Dry Fly	Steelhead/ Salmon	4XF	S	TU-L	DLim	Blk		Chem	Gaelic Supreme Salmon Single Dry Fly	Forged.	2-10

GAELIC SUPREME Characteristics

Hook	Use	Wire	Shank	Eye	Bend	Finish	Barb	Point	Common Name	Remarks	Size
Single Low Water	Steelhead/ Salmon	1XH	2XL	TU-L	DLim	Blk		Chem	Gaelic Supreme Salmon Single Low Water	Forged.	2/0-12
Single Wet	Steelhead/ Salmon	2XH	S	TU-L	DLim	Blk		Chem	Gaelic Supreme Salmon Single Wet	Forged.	2/0-12
Steelhead Egg Hook	Egg	3XH	1XS	S	N.I.-M	Blk		Chem	Gaelic Supreme Salmon Steelhead Egg Hook	Mfg. states bend as "Salmon Bend." Forged.	4-12
Treble Low Water	Treble	2XF	2XL	TU-L	N.I.-Treb	Blk	Out	Chem	Gaelic Supreme Salmon Treble Low Water	Forged.	4-12
Treble Wet	Treble	1XH	2XL	TU-L	N.I.-Treb	Blk	Out	Chem	Gaelic Supreme Salmon Treble Wet	Forged.	4-12
Trout Streamer	Streamer	3XH	5XL	TU-M	P	Nick	Mini	Chem	Gaelic Supreme Trout Streamer Hook	Forged.	2-12
Wet Fly Hook	Wet	S	S	TU-M	Spr	Olive	Mini	Chem	Gaelic Supreme Trout Wet Fly Hook	Eye partially TU. Forged.	6-16
Wide Gape Midge	Midge/Pupae	S	S	TU-M	P	M-Nick	Mini	Chem	Gaelic Supreme Trout Wide Gape Midge	Eye partially TU. Matte finish. Wire sized so as not to tear fish's mouth. Forged.	20-24

Note: F-F means "Francis-Francis"

Gamakatsu	Daiichi	Dai-Riki	Eagle Claw	Gaelic Supreme	Hayabusa	Heritage	Kamasan	Mustad	Orvis	Partridge	TMC	Turrall	Varivas	VMC
C11-5L2H[6]	1730[1]	700-B					B810	81002BR						
C12	1130[1,3]	135[3]						80250BR[1,3]			2487[1,3]	Grub		9902[3]
C12U			L144[2]	Caddis[3,4]	***388***[5]			37160[1,3] 37161[3,4] 37162[3,4]			***205BL***[1,3,5]			
C13U							B420[4]			Z11[3]				
C14S	1510[1]							9479[1,3]			105[1]			
L10-2H	1710[1,3] 1550	075	057					3906[1] 3908[1]		G3A SH1[3,4]	3769[1]		977[3,4]	
L11S-3H	2451[1,3,4]							3406[1,3]			800S[1]			
P-10	1170[1] 1180 ***1190*** 1182[4]	305 300	L059[2]			Trout Down Eye[3]		94840[1]	1509[3] ***1877***[3] 4864[3,5]	L2A[3] GRS2A[3,4]	9300[1] 5210[1]	Sproat[3]		9060 9288
P10-2L1H	1280[1] 1710	730	L063[2]				B200[3]	94831[1]			5212[1]			9279
R10-BN		305[3,4]				Trout Down Eye[3,4]						Sproat[3,4]		
Russian River	2340[3]							36717[1,3]			300[3]			
S10	1170[1] 1180 ***1190*** 1182[4] 1550[3,4]	305[4]	L057[3,4]			Trout Down Eye[3,4]		94840[1,4] AC94840[4] ***94845***[4] 3906[3,4] 3399[3,4] 3399A[3,4] 3399N[3,4]	1509[3,4] ***1877***[4,5] 4864[3,4]	G3A[3,4] J1A[3,4] CS7MW[3]	5210[1] 100[3,4] ***100BL***[3,4]	Sproat[3,4] ***Barbless Sproat***[3,4] Bead Standard Nymph[3,4]		9060[3] 8526[3,4]
S10-2S	1530[1]							948381[3,4]			921[1]			

Hooks that are ***bold italicized*** are barbless.
[1]Gamakatsu considers this hook to be similar. The brand at the top of the column may or may not concur.
[2]The brand at the top of the column considers the Gamakatsu hook to the left to be similar. Gamakatsu may or may not concur.
[3]Similar in all aspect except for the bend. See the corresponding characteristic tables.
[4]Similar in all aspects except the finish. See the corresponding characteristic tables.
[5]Similar in all aspects except the point position. See the corresponding characteristic tables.
[6]Swimming nymph hooks. Weight and length varies. Eye, Bend, finish and shape are similar unless otherwise noted.
[7]The Trey Combs Big Game hook is unique and manufactured by Gamakatsu. A possible Gamakatsu substitution is the SL11-3H. The Trey Combs Big Game hook is exclusively distributed by Spirit River, Inc. at 423 Winchester St., Roseburg, OR 97470, (541) 440-6916.

Gamakatsu	Daiichi	Dai-Riki	Eagle Claw	Gaelic Supreme	Hayabusa	Heritage	Kamasan	Mustad	Orvis	Partridge	TMC	Turrall	Varivas	VMC
S10-3F	1100[1,4]		L059[4] L060[3,4] ***L061B***[4]		572[3]		B400[3,4]	94833[1] 94840[4] AC94840[4] ***94845***[4] 80000BR[4] ***80010BR***[4]	1523[4]	L3A[3,4] ***L3AY***[3,4] GRS3A[3,4]	5230[1] 5210[4] 102Y[3] ***103BL***[3] ***900BL***[3,4,5] 100[3,4] ***100BL***[3,4] 146Z			9280[3,4] V7061[4]
S10-B	***1190***[1,4] 1170[4] 1180[4] 1182[4]	305[4]	L057[3,4]			Trout Down Eye[3,4]		***94845***[1]	1509[3,4] ***1877***[4,5] 4864[3,4]	G3A[3,4] J1A[3,4] CS7MW[5]	5210[1] ***902BL***[3] ***103BL***[3] ***900BL***[3,5]	Sproat[3,4]		9060[4]
S10S	1110[1,3]							94859[1,3]			101[1,3]			
S10U	1330[1,4]			Wet Fly[3,4]				94842[1,4]			500U[1,4]			9281[3,4]
S11-4L2H	2220[1,3]	700[3]	L281[2,3]		***382***[3] 385[3]			33960[1] 79580[3]			5263[1,3]			
S11S-4L2H	1750[1]							9674[1]			9394[1]			
S12S-1F	1640[1,3] 1110[3]	310[3]						3366[1] 94859[3]	170T[3,4]		101[3]			
S-13S-M	1480[1,3,4] 1110[4]							94859[1,4]	170T[3]		501[1,4] 101[4]		988[3]	
S14S-3H	2451[1,3,4]							3406[1,3]			800B[1,3,4,5] 800S[1,3,4,5]			
SC15	3111[3,4]		L067[3,4] 254SS[3,4] L2454N[3,4] 254N[3,4] 254[3,4] 254BK[3,4] L253[3,4]							MM3ST[3,4] Z12[3,4] Z1[3,4] ***Z1Y***[3,4] GRS 1Z[3,4] Z2[3,4] Z10[3,4]		Salt Water[3,4]		7250NI[3,4] 8410[3,4]
SL11-3H[7]											800S[1]			
SP11-3L3H										CS53[3,4]				

Gamakatsu	Daiichi	Dai-Riki	Eagle Claw	Gaelic Supreme	Hayabusa	Heritage	Kamasan	Mustad	Orvis	Partridge	TMC	Turrall	Varivas	VMC
SS15	2546[1,3,4]							31014[3,4] 34007[3,4] 3407[3] 3366[3,4] 34009[3,4] 3406[3,4] 3406B[3,4]	9034[3,4]	CS51 CS41[3,4]	811S[1,2,3]	Salt Water[3,4]		9752[3,4]
T10-3H	2421[1,3]							90240[1]			7989[1,3]			
T10-6H	2441[1] 2050[3] 2051[3] 2052[3,4] 2055[3,4] 2059[3] 2131[3,4] 2135[3,4] 2139[3]							36890[1] 90240		O1[3]	7999[1,3] 7989[3]			
Trey Combs Big Game[7]								3407SS[3,4,5]		CS52[3,4,5] Z8[3,4,5]	800S[3,4,5]			

Hooks that are ***bold italicized*** are barbless.
[1]Gamakatsu considers this hook to be similar. The brand at the top of the column may or may not concur.
[2]The brand at the top of the column considers the Gamakatsu hook to the left to be similar. Gamakatsu may or may not concur.
[3]Similar in all aspect except the bend. See the corresponding characteristic tables.
[4]Similar in all aspects except the finish. See the corresponding characteristic tables.
[5]Similar in all aspects except the point position. See the corresponding characteristic tables.
[6]Swimming nymph hooks. Weight and length varies. Eye, Bend, finish and shape are similar unless otherwise noted.
[7]The Trey Combs Big Game hook is unique and manufactured by Gamakatsu. A possible Gamakatsu substitution is the SL11-3H. The Trey Combs Big Game hook is exclusively distributed by Spirit River, Inc. at 423 Winchester St., Roseburg, OR 97470, (541) 440-6916.

GAMAKATSU Characteristics

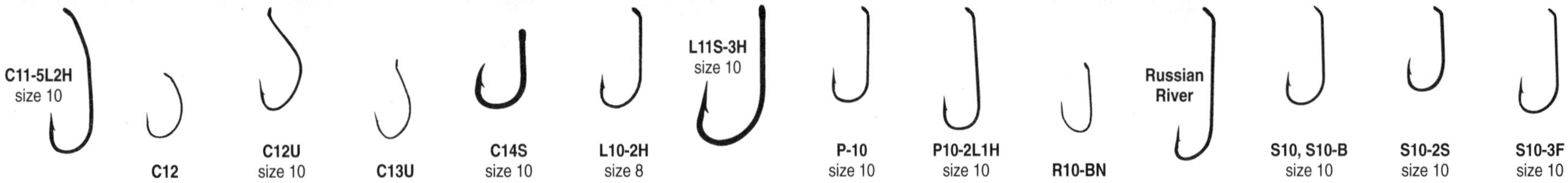

Hook	Use	Wire	Shank	Eye	Bend	Finish	Barb	Point	Common Name	Remarks	Size
C11-5L2H	Nymph	2XH	5XL-BD	TD	P	Bron			Stone Fly 5X Long, 2X Strong	Stoneflies, crab patterns.	6-12
C12	Scud/Shrimp	S	S-C	TD	C-WG	Bron			Down Eye Scud		10-20
C12U	Caddis	S	S-C	TU	C-WG	Bron			Caddis	Pupae.	6-16
C13U	Dry	1XF	S-C	TU-M	C-WG	Blk			Keel Balance	Matte blk finish. Eye slightly TU.	10-20
C14S	Egg	2XH	4XS	S	P	Blk			Glo Bug		2-12
L10-2H	Nymph	2XH	S	TD	Spr	Bron			Standard Nymph, 2X Strong	Wet flies.	6-12
L11S-3H	Steelhead/Salmon	3XH	S	S	Lim	Bron			Salmon and Steelhead, 3X Strong		2-8
P-10	Dry	S	S	TD	P	Bron			Traditional Dry		4-22
P10-2L1H	Dry	1XH	2XL	TD	P	Bron			Large Dry, 2X Long, 1X Strong	Nymphs, wet flies.	4-16
R10-BN	Dry	S	S	TD	N.I.-M	Blk	No Barb		Retainer Bend-Barbless	Modified bend. Matte blk finish.	10-20
Russian River	Streamer	1XH	6XL	TD	Osh	Bron			Russian River Fly	Meets Russian River regulations.	2
S10	Dry	S	S	TD	P	Blk			Standard Down eye	Wet flies, nymphs. Same as S10-B.	2-24
S10-2S	Nymph	S	2XS	TD	P	Blk			Small Nymph, 2X Short	Emergers.	6-18
S10-3F	Dry	1XF	S	TD	P	Blk			Standard Dry Fly, 1X Fine		4-20
S10-B	Dry	S	S	TD	P	Blk	No Barb		Standard Down Eye, Barbless	Wet flies, nymphs. Same as S10.	6-16

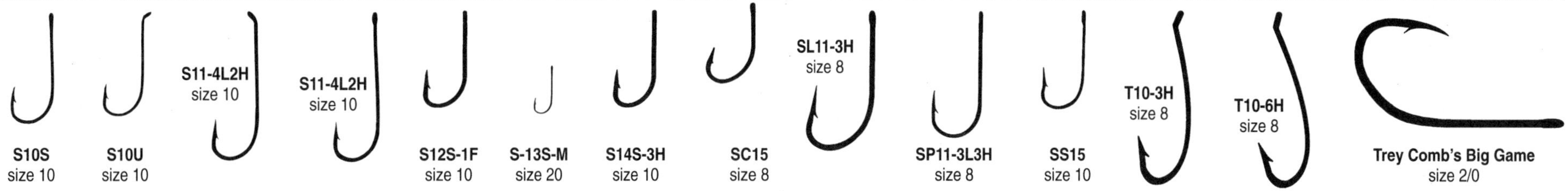

Hook	Use	Wire	Shank	Eye	Bend	Finish	Barb	Point	Common Name	Remarks	Size
S10S	Dry	S	S	S	P	Blk			Straight Eye Dry		4-14
S10U	Dry	S	S	TU	P	Blk			Up-Eye Dry		4-14
S11-4L2H	Streamer	2XH	4XL	TD	Spr	Bron			Streamer, 4X Long, 2X Strong	Large wet flies.	6-12
S11S-4L2H	Streamer	2XH	4XL	S	Spr	Bron			Strait Eye Streamer, 4X Long, 2X Strong.	Wet flies, Woolly Buggers.	6-12
S12S-1F	Dry	1XF	S	S	Spr	Bron			Multi Use, 1X Fine	Caddis, wet flies.	8-18
S-13S-M	Midge/Pupae	1XF	S	S	P	Blk			Midge, 1X fine	Tricos, variants.	20-24
S14S-3H	Nymph	3XH	S	S	Spr	Bron			Multi Use, 3X Strong	Wet flies.	8-12
SC15	Saltwater	S	S	S	WG	Tin			Saltwater Wide Gape	Shrimp patterns.	2/0-8
SL11-3H	Saltwater	3XH	S	S	Lim	Tin			Saltwater, 3X Strong	Bonefish, redfish, tarpon, billfish.	8/0-8
SP11-3L3H	Saltwater	3XH	3XL	S	P	Tin			Saltwater Perfect Bend	Bonefish, tarpon.	6/0-8
SS15	Saltwater	S	S	S	P	Tin			Standard Saltwater	Bonefish, redfish.	4/0-18
T10-3H	Steelhead/Salmon	1XF	S-C	TU	Lim	Blk			Salmon Dry, 1X Fine		2-8
T10-6H	Steelhead/Salmon	S	S-C	TU-L	Lim	(Rmks)			Salmon, Tapered Loop, Up-Eye	Available in blk, red, blue, & grn.	2/0-8
Trey Combs Big Game	Saltwater	3XH	S	S-O	Spr-M	Tin	Mini	Chem		Designed by Trey Combs. Eye designed to handle large shock tippets.	10/0-2

Hayabusa	Daiichi	Dai-Riki	Eagle Claw	Gaelic Supreme	Gamakatsu	Heritage	Kamasan	Mustad	Orvis	Partridge	TMC	Turrall	Varivas	VMC
371	1150[3] 1155[3,4]							81001BR		K2B		Sedge		
372	1170 1180 ***1190*** 1182[4]	305					B401[3] ***B402*** K40		1509[3] ***1877***[5] 4864[3]				980	9060
373	1550 1170[3] 1180[3] ***1190***[3] 1182[3,4]	305[3]	L057					3906	1509[3] ***1877***[3,5]	G3A		Sproat ***Barbless Sproat***	980[34] 98r[3,4]	
375	1330[3]						B440							
376	1710	730	L063					9671 AC9671	1524		5262			9279[3]
377		310							4641[3]					
378	1170 1180 ***1190*** 1182[4]	305					***B402*** K40 B401[3]		1509[3] ***1877***[5] 4864[3]				980	9060
379		310							4641[3]					
380								80150BR	1512	K6ST[3]	400T			
381	1710	730	L063					9671 AC9671	1524		5262			9279[3]
382	2220[3]				S11-4L2H[3]			33960[3]			5263[3]			
383	1270[3] 1273[3,4]		L052[3]						1510[3]		200R[3,5] ***200RBL***[3,5]			
384														
385	2220[3]				S11-4L2H[3]			33960[3]			5263[3]			
386	1170 1180 ***1190*** 1182[4]	305					***B402*** K40 B401[3]		1509[3] ***1877***[5] 4864[3]				980	9060

Hayabusa	Daiichi	Dai-Riki	Eagle Claw	Gaelic Supreme	Gamakatsu	Heritage	Kamasan	Mustad	Orvis	Partridge	TMC	Turrall	Varivas	VMC
387		135[3]		Grub[3,4]		Grub[3]	B110[3]	80200BR[3]			2457[3]			
388			L144[4]	Caddis[4,5]	C12U[5]			37160[3,5] 37161[3,4,5] 37162[3,4,5]			***205BL***[3]			
552	1280	730								H1A[3]	5212			V7062
572	1100[3]		L060 L059[3] ***L061B***[3]		S10-3F[3,4]		B400[3]	80000BR[3] ***80010BR***[3] 94840[3] AC94840[3] ***94845***[3]	1523[3]	L3A[3] ***L3AY***[3] GRS3A[3,4]	100[3] ***100BL***[3] 102Y[3,4] ***103BL***[3] 146Z[3] 5210[3] ***900BL***[3,5]			9280[3] V7061[3]
751							B405[3]	94838[3]		E6A ***CS20***[3,4]	921		981[3]	9282[3]
752								94838[3]			5230[3]		981[3]	9288[3]
753						Stronghold[3]						Stronghold[3]	973[3]	
761														
762														
771	2421[3]						B180	36890[3]				Salmon Single[3]		

Hooks that are ***bold italicized*** are barbless.
[1]Hayabusa considers this hook to be similar. The brand at the top of the column may or may not concur.
[2]The brand at the top of the column considers the Hayabusa hook to the left to be similar. Hayabusa may or may not concur.
[3]Similar in all aspects except the bend. See the corresponding characteristic tables.
[4]Similar in all aspects except the finish. See the corresponding characteristic tables.
[5]Similar in all aspects except the point position. See the corresponding characteristic tables.

HAYABUSA Characteristics

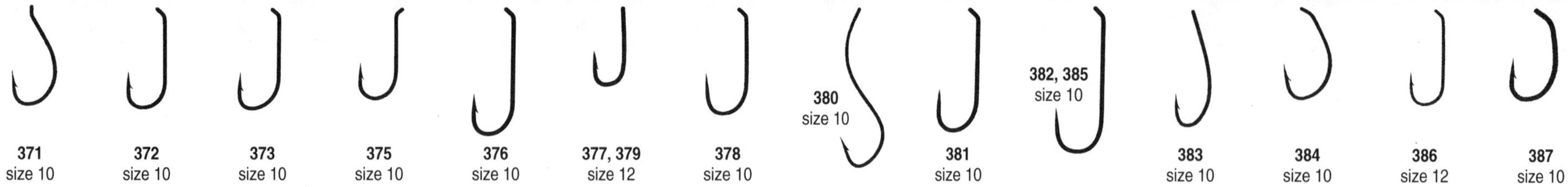

Hook	Use	Wire	Shank	Eye	Bend	Finish	Barb	Point	Remarks	Size
371	Scud/Shrimp	S	S-C	TU	C-WG	(Rmks)	Mini	Chem	Emergers, pupae. Finish available in bron & blk nick. Forged.	8-16
372	Dry	S	S	TD	P	(Rmks)	Mini	Chem	Finish available in bron & blk nick. Forged.	8-24
373	Dry	S	S	TD	Spr	(Rmks)		Chem	Wet flies & nymphs. Finish available in bron & blk nick.	4-20
375	Dry	S	1XS	TU	Spr	(Rmks)		Chem	Finish available in bron & blk nick. Forged.	8-22
376	Nymph	2XH	2XL	TD	P	(Rmks)		Chem	Streamers. Finish available in bron & blk nick. Forged.	4-18
377	Dry	S	S	S-O	P	(Rmks)		Chem	Midges. Same as 379. Finish available in bron & blk nick.	12-24
378	Dry	S	S	TD	P	(Rmks)	No Barb	Chem	Finish available in bron & blk nick.	8-24
379	Dry	S	S	S-O	P	(Rmks)	No Barb	Chem	Midges. Same as 377. Finish available in bron & blk nick.	16-24
380	Nymph	S	5XL-BU	S	Spr-M	(Rmks)	Mini	Chem	Swimming Nymphs. Finish available in bron & blk nick. Forged.	8-14
381	Nymph	2XH	2XL	TD	P	(Rmks)	No Barb	Chem	Small Streamers. Finish available in bron & blk nick.	10-18
382	Streamer	2XH	4XL	TD	P-WG	(Rmks)	No Barb	Chem	Large Nymphs. Same as 385. Finish available in bron & blk nick. Mfg. states bend is 2X wide.	8-18
383	Nymph	1XH	2XL-C	S	York	(Rmks)		Chem	Wet flies. Slightly humped shank. Finish available in bron & blk nick.	8-18
384	Scud/Shrimp	S	S	TD	C-WG	(Rmks)	Mini	Chem	Finish available in bron & blk nick. Forged.	10-18
385	Streamer	2XH	4XL	TD	P-WG	(Rmks)	Mini	Chem	Large Nymphs. Same as 382. Finish available in bron & blk nick. Mfg. states bend is 2X wide.	8-18
386	Dry	S	S	TD	P	(Rmks)	Mini	Chem	Finish available in bron & blk nick.	12, 14
387	Scud/Shrimp	3XH	2XS-C	TD	C-WG	(Rmks)		Chem	Finish available in bron & blk nick. Mfg. states bend is 2X wide. Forged.	6-18

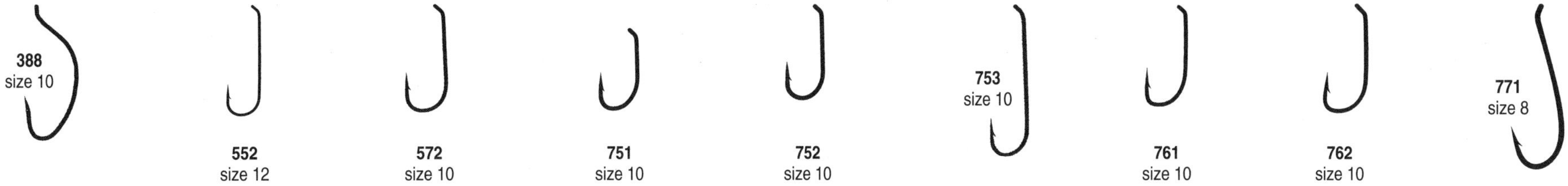

Hook	Use	Wire	Shank	Eye	Bend	Finish	Barb	Point	Remarks	Size
388	Caddis	S	S-C	TU	C-WG	(Rmks)	No Barb	SDP-C	Shrimp. Finish available in bron & blk nick. Mgf. states bend is 4X wide. Forged.	8-20
552	Dry	1XF	2XL	TD	P	(Rmks)		Chem	Terrestrials. Finish available in bron & blk nick. Forged.	6-16
572	Dry	1XF	S	TD	Spr	(Rmks)		Chem	Terrestrials. Finish available in bron & blk nick. Forged.	8-24
751	Dry	1XF	3XS	TD	Spr	(Rmks)		Chem	Shrimp. Finish available in bron & blk nick.	6-16
752	Dry	2XF	1XS	TD	Spr	(Rmks)		Chem	Finish available in bron & blk nick. Forged.	10-20
753	Streamer	3XH	4XL	TD	P	(Rmks)		Chem	Large Nymphs. Finish available in bron & blk nick. Forged.	4-14
761	Wet	3XH	S	TD	Spr	(Rmks)		Chem	Nymphs. Finish available in bron & blk nick. Forged.	6-16
762	Nymph	3XH	S	TD	Spr	(Rmks)		Chem	Wet flies. Shank is slightly longer than the 761. Finish available in bron & blk nick. Forged.	8-16
771	Steelhead/ Salmon	S	S	TU-L	York	Blk		Chem		2/0-12

HERITAGE Cross-Reference

Heritage	Dai-Riki	Gaelic Supreme	Gamakatsu	Hayabusa	Kamasan	Mustad	Orvis	Partridge	TMC	Turrall	Varivas
Allcock Streamer		MMRS-6XL				33957	1511[3]	D7A[3]			
Grub Hook	135[3]	Grub[4]		367[3]	B110[3]	80200BR[3]			2457[3]		
Harrison-Bartleet Pattern		F-F Dee Style Blind Eye									
Nymph Hook		Nymph									
Pryce-Tannant Pattern		F-F Light Blind Eye						CS6		Salmon Blind Eye	
Single Salmon		F-F Single Wet									
Single Salmon Low Water		F-F Low Water				80500BL[3]				Salmon Low-water[3]	
Stronghold				753[3]						Stronghold	973[3]
Trout Down Eye			P-10[3] ***R10-BN***[3,4] S10[3,4] ***S10-B***[3,4]							Sproat	
Trout Longshank	710					9672 AC9672 38941[3]	1526	SH3[3,4]		Extra-long Streamer	
Trout Up Eye										Up-eye Dry Fly	
Trout Xtra Fine								***CS2001***[5] CS21[3,4] ***E3AY***[3] E4A[3] GRS3A[3,4] L3A[3] ***L3AY***[3,4]		***Barbless Extra-fine*** Extra-fine Dry Fly	

Hooks that are ***bold italicized*** are barbless.
[1]Heritage considers this hook to be similar. The brand at the top of the column may or may not concur.
[2]The brand at the top of the column considers the Heritage hook to the left to be similar. Heritage may or may not concur.
[3]Similar in all aspects except the bend. See the corresponding characteristic tables.
[4]Similar in all aspects except the finish. See the corresponding characteristic tables.
[5]Similar in all aspects except the point position. See the corresponding characteristic tables.

Hook	Use	Wire	Shank	Eye	Bend	Finish	Barb	Point	Remarks	Size
Allcock Streamer	Streamer	S	6XL	TD-M	Spr-M	Bron			Saltwater. Double-coated finish gives the hook a "nearly olive" color.	2 & 4
Grub Hook	Scud/Shrimp	1XH	S-C	TD	C	Bron	Mini	Chem	Grubs. Forged.	8-14
Harrison-Bartleet Pattern	Display	S	5XL	No	DLim	Blk			Handmade. Longer shanks in each size available. Mfg. states these are exact duplicates of original antique hooks.	8/0-6/0
Nymph Hook	Nymph	S	3XL	TU-M	P	Olive	Mini	Chem	Eye partially TD. Forged.	10-16
Pryce-Tannant Pattern	Display	2XF	1XL	No	DLim	Blk			Handmade. Longer shanks in each size available.	6/0-2/0

HERITAGE Characteristics

Single Salmon
size 4

Single Salmon Low Water
size 4

Stronghold
size 6

Trout Down Eye
size 12

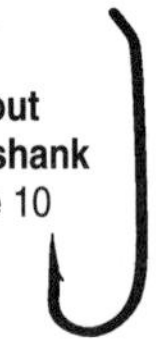

Trout Longshank
size 10

Trout Up Eye
size 10

Trout Xtra Fine
size 12

Hook	Use	Wire	Shank	Eye	Bend	Finish	Barb	Point	Remarks	Size
Single Salmon	Steelhead/Salmon	2XH	S	TU-L	DLim	Blk		Chem	Forged.	2-10
Single Salmon Low Water	Steelhead/Salmon	S	2XL	TU-L	DLim	Blk		Chem	Forged.	4-10
Stronghold	Streamer	3XH	3XL	TD	Spr	Bron	Mini	Chem	Long-bodied nymphs. Forged.	2-8
Trout Down Eye	Nymph	S	S	TD	Spr	Bron	Mini	Chem	Wet flies. Forged.	8-14
Trout Longshank	Streamer	S	3XL	TD	P	Bron	Mini	Chem	Forged.	4-10
Trout Up Eye	Dry	3XF	S	TU	Spr-R	Bron	Mini	Chem	Forged.	10-16
Trout Xtra Fine	Dry	4XF	S	TD	P	Bron	Mini	Chem	Forged.	10-16

Kamasan	Daiichi	Dai-Riki	Eagle Claw	Gaelic Supreme	Gamakatsu	Hayabusa	Heritage	Mustad	Orvis	Owner	Partridge	TMC	Turrall	Varivas	VMC
B100								80250BR[1,3] ***80260BR***[3]			K4A[3]	2487[1,3] 2487BL[3] 2487G[3]			
B100G								80250BR[3,4] ***80260BR***[3,4]			K4A[4]				
B100N								80250BR[3,4] ***80260BR***[3,4]			K4A[4]				
B110		135[3]		Grub[3]		387[3]	Grub[3,4]	80200BR[3]				2457[1,3]			
B160	1510[3]														
B170		300[3]						3906 AC3906B 7957B[3]					Extra-long Nymph[3] Bead Extra-long Nymph[3]		8527
B175	1550[1]		L1197B NT1197[4] L1197[4] L1197N[4]					3906[1]	167T			3769[1] 9300[3] 700[3,4,5]			
B180	2421[1,3]					771		36890[3]				7999[1,3]	Salmon Single[3]		
B190	2441[1,3]			Single Low Water[3]				36890[1,3]	1645[3]		M[3]	7999[1,3]		974[3]	
B200	1710[3]				P10-2L1H[3]										9279[3]
B220	1270[1,4] 2151[1,3]							80050BR[1,3] ***80060BR***[3]			K12ST[3,4] GRS12ST[3,4]	200R[1,3,5] ***200RBL***[3,5]			
B270											R1A[3] R2A[3]				
B280				Double Low Water[3] Double Wet[3]							Q1[3] GRSQ2[3,4]		Salmon Double Low-water[3] Salmon Double[3] Sproat Double[4]		9410[3,4]
B380				Treble Low Water[3] Treble Wet[3]							X2B[3,5] GRS X2B[3,4,5] CS12[3]		Salmon Treble[3,5] Salmon Treble Low-water[3,4,5]		

Hooks that are ***bold italicized*** are barbless.
[1]Kamasan considers this hook to be similar. The brand at the top of the column may or may not concur.
[3]Similar in all aspects except for the bend. See the corresponding characteristic tables.
[4]Similar in all aspects except the finish. See the corresponding characteristic tables.
[5]Similar in all aspects except the point position. See the corresponding characteristic tables.
[6]Weight and length varies. Eye, bend, finish and shape are similar unless otherwise noted.

Kamasan	Daiichi	Dai-Riki	Eagle Claw	Gaelic Supreme	Gamakatsu	Hayabusa	Heritage	Mustad	Orvis	Owner	Partridge	TMC	Turrall	Varivas	VMC
B400	1100[3]		L059[3] L060[3] ***L061B***[3]		S10-3F[3,4]			80000BR[3] ***80010BR***[3] 94840[3] AC94840[3] ***94845***[3]	1523[3]		L3A[3] ***L3AY***[3] GRS3A[3,4]	100 ***100BL*** 102Y[3,4] ***103BL***[4] 146Z[4] 5210[3,4] ***900BL***[3,4,5]			9280 V7061[3]
B401	1100[1,3] 1170[3] 1180[3] ***1190***[3] 1182[3,4]	305[3]				372[3] 373[3] ***378***[3] 386[3]		94840[1,3] AC94840[3] ***94845***[3]	1509[3] ***1877***[3,5] 4864[3]			100[1]		980[3]	9060[3]
B402	***1190***[1] 1170 1180 1182[4]	305				372 373[3] ***378*** 386		***94845***[1] 94840 AC94840	1509[3] ***1877***[5] 4864[3]			***900BL***[1,3,4,5]		980	9060
B405						751[3]		94838			***CS20***[3,4,5] E6A[3]	921[3]		981[3,4]	9282[3] ***9388***[3]
B410								94859[1,3]				101[1,3]			
B420	1150[1,3] 1155[3,4]				C13U[3,4]						Z11[3,4]				
B440	1330[1]					375[3]		94842[1]				500U[1]			
B800								79580[1,3] 33960[3] 3498AD[3,4]	8808[3]		CS11/1[3,4] CS11/2[3,4] CS11GRS[3,4] ***CS2002***[3,5]				9283[3]
B810[6]	1730	700-B			C11-5L2H			81002BR							
B820	1750							9674[1] 9674A[4] 9674B[4]	0167						
B830			L063					9671[1] AC9671	1524 1638[3]						

Kamasan	Daiichi	Dai-Riki	Eagle Claw	Gaelic Supreme	Gamakatsu	Hayabusa	Heritage	Mustad	Orvis	Owner	Partridge	TMC	Turrall	Varivas	VMC
B940	2461 2460[3] 2462[3] 1750[3,4]							3261[4] 3262[4] 80400BLN[4] 80400D[3,4]		5320[3,4]	D3ST[3,4]				9255[3,4]
B990								35647					Tube Fly-Treble[4]		
K18								94842[3]							9281[3]
K40	1170 1180 ***1190*** 1182[4]	305				372 ***378*** 386 373[3]			1509[3] ***1877***[5] 4864[3]					980	9060
K80								79580 33960[3] 3498AD[3,4]	8808		CS11/1[3,4] CS11/2[3,4] CS11GRS[3,4] ***CS2002***[5]				9283
K93	X452[3,4]		250[3,4]						XPoint[3,4]						7255[3,4]

Hooks that are ***bold italicized*** are barbless.
[1]Kamasan considers this hook to be similar. The brand at the top of the column may or may not concur.
[3]Similar in all aspects except for the bend. See the corresponding characteristic tables.
[4]Similar in all aspects except the finish. See the corresponding characteristic tables.
[5]Similar in all aspects except the point position. See the corresponding characteristic tables.
[6]Weight and length varies. Eye, bend, finish and shape are similar unless otherwise noted.

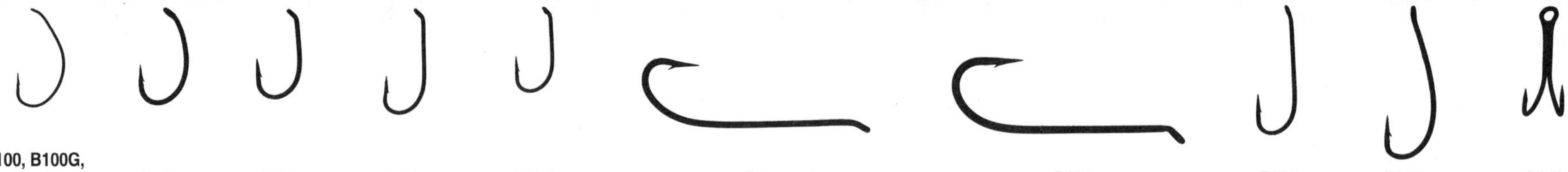

Hook	Use	Wire	Shank	Eye	Bend	Finish	Barb	Point	Common Name	Remarks	Size
B100	Scud/Shrimp	S	1XS-C	TD	C-WG	Bron		Chem		Same as B100G, B100N. Grubs. Forged.	10-16
B100G	Scud/Shrimp	S	1XS-C	TD	C-WG	Gold		Chem		Same as B100, B100N. Grubs. Forged.	10-16
B100N	Scud/Shrimp	S	1XS-C	TD	C-WG	Nick		Chem		Same as B100, B100G. Grubs. Forged.	10-16
B110	Scud/Shrimp	1XH	1XS-C	TD	C-WG	Bron		Chem		Grubs. Forged.	10-16
B160	Nymph	S	3XS	TD	Spr-M	Bron		Chem		Forged.	6-16
B170	Nymph	S	1XL	TD	Spr	Bron		Chem		Wet flies. Forged.	2-16
B175	Nymph	1XH	S	TD	Spr	Bron		Chem		Wet flies. Forged.	2-16
B180	Steelhead/Salmon	S	S	TU-L	York	Blk		Chem		Low water. Forged.	4-12
B190	Steelhead/Salmon	1XH	S	TU-L	York	Blk		Chem		Forged.	2/0-2
B200	Nymph	1XH	2XL	TD	Spr	Bron		Chem	Deepwater Nymph	Steelhead flies. Forged.	6-16
B220	Streamer	S	3XL	S	York	Blk-Nick		Chem		Stoneflies, terrestrials.	6-14
B270	Double	1XH	S	TD-L	Spr-Dbl	Bron		Chem		Wet flies, streamers. Forged.	6-14

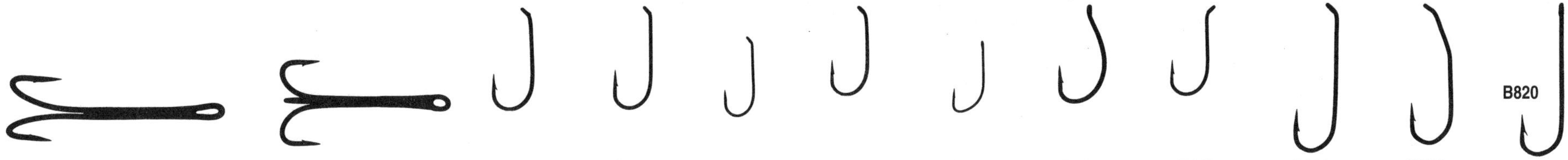

Hook	Use	Wire	Shank	Eye	Bend	Finish	Barb	Point	Common Name	Remarks	Size
B280	Double	S	S	TU-L	Spr-Dbl	Blk		Chem		Low-water. Forged.	4-12
B380	Treble	S	1XL	TU-L	Spr	Blk		Chem		Low-water. Forged.	4-16
B400	Dry	1XF	S	TD	P-M	Bron		Chem		Emergers. Forged.	10-16
B401	Dry	S	S	TD	P-M	Bron	Mini	Chem		Forged.	8-20
B402	Dry	S	S	TD	P	Bron	No Barb	Chem		Forged.	14-20
B405	Dry	1XF	1XS	TD	P-M	Bron		Chem		Forged.	8-20
B410	Midge/Pupae	2XF	1XS	S	Lim-M	Bron		Chem		Dry flies, Tricos. Crystal bend.	12-22
B420	Caddis	1XF	S-C	TU-M	C-WG-M	Bron		Chem		Eye partially turned. Sedges, emergers, parachutes.	8-16
B440	Dry	S	1XS	TU	P	Bron		Chem		Forged.	8-20
B800	Streamer	S	4XL	TD	P-M	Bron		Chem		Long-bodied nymphs. Forged.	1-14
B810	Nymph	S	4XL-BD	TD	P-M	Bron		Chem		Caddis larvae. Forged.	6-14
B820	Streamer	S	4XL	S	P	Bron		Chem		Zonkers, Woolly Buggers. Forged.	4-12

KAMASAN — Characteristics

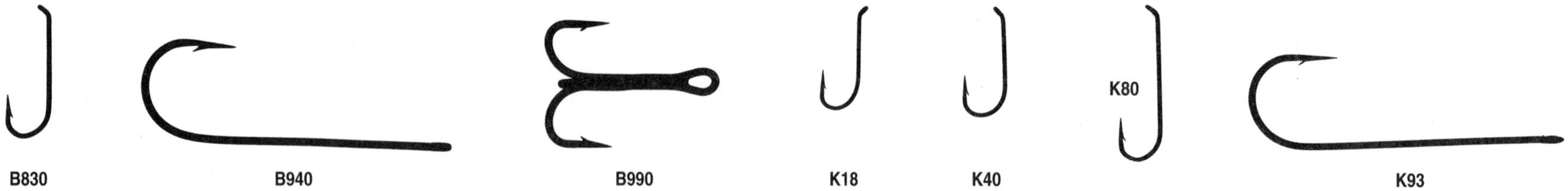

Hook	Use	Wire	Shank	Eye	Bend	Finish	Barb	Point	Common Name	Remarks	Size
B830	Streamer	S	2XL	TD	P	Bron		Chem		Nymphs, terrestrials, beadheads. Forged.	2-16
B940	Saltwater	1XH	4XL	S	Aber	Blk	Mini	Chem		Musky, bass, pike.	6/0-6
B990	Treble	1XH	S	S-O	P-Treb	Bron		Chem		Tube flies.	2-14
K18	Dry	1XF	S	TU	Spr	Bron	Mini				10-18
K40	Dry	S	S	TD	P	Bron	Mini				8-16
K80	Streamer	S	4XL	TD	P	Bron	Mini				6-14
K93	Bass/Pike	S	1XL	S	Aber	Nick	Mini			Saltwater.	4/0-6

Mustad	Daiichi	Dai-Riki	Eagle Claw	Gaelic Supreme	Gamakatsu	Hayabusa	Heritage	Kamasan	Orvis	Owner	Partridge	TMC	Turrall	Varivas	VMC
31014	2546[3]				SS15[3,4]				9034[3]		CS51[3,4] CS41[3,4]		Salt Water[3]		9752[3,4]
3257B Discontinued	1170[3] 1180[3] ***1190***[3] 1182[3,4]	305[3]				373 372[3] ***378***[3] 386[3]		B401[3] ***B402***[3] K40[3]	1509[3] ***1877***[3,5] 4864[3]					980[3]	9060[3]
3261	2460[2] 2461[3] 2462[3]							B940[4]		5320[3,4]					
3262	2461[2,4] 2460[4] 2462[4]							B940[4]		5320[3,4]					
3366	2546[3,4]				S12S-1F[2] S10S[3,4] SS15[3,4]				9034[3,4]		CS41[3,4] CS51[3,4]		Salt Water[3,4]		9752[3,4]
33900			L200[3]									511S[3,4]			
33903			L200[3]									511S[3,4]			
33956															
33957			L281[3]	MMRS-6XL			Allcock Streamer		1511[3]		D7A[3]				
33960		700[3]			S11-4L2H[2,3]			B800[3] K80[3]	8088[3]		CS11/1[3,4] CS11/2[3,4] CS11GRS[3,4] ***CS2002***[3,5]				9283[3]
3399	1550	305[3]	L060[2] L057		S10[3,4]	373					G3A J1A[3] CS7MW[3,4]		Sproat ***Barbless Sproat*** Bead Standard Nymph[3]	98r[3,4]	8526
3399A	1550	305[3]	L057		S10[3,4]	373					G3A J1A CS7MW[3,4]		Sproat ***Barbless Sproat*** Bead Standard Nymph[3]	98r[3,4]	8526
3399D			L059								***CS27GRS***[3,4] GRS2A[3,4] L2A[3]	***902BL***[3,4]		98i[3,4]	V7060[3]

Hooks that are ***bold italicized*** are barbless.

[1]Mustad considers this hook to be similar. The brand at the top of the column may or may not concur.

[2]The brand at the top of the column considers the Mustad hook to the left to be similar. Mustad may or may not concur.

[3]Similar in all aspects except the bend. See the corresponding characteristic tables.

[4]Similar in all aspects except the finish. See the corresponding characteristic tables.

[5]Similar in all aspects except the point position. See the corresponding characteristic tables.

MUSTAD Cross-Reference

Mustad	Daiichi	Dai-Riki	Eagle Claw	Gaelic Supreme	Gamakatsu	Hayabusa	Heritage	Kamasan	Orvis	Owner	Partridge	TMC	Turrall	Varivas	VMC
3399N	1550[4]	305[3,4]	L057[4]		S10[3,4]	373[4]					G3A[4] J1A[3,4] CS7MW[3,4]		Sproat[4] ***Barbless Sproat***[4] Bead Standard Nymph[3,4]	98r[3,4]	8526[4]
34005															
34007	2546[2]	930[2]	254SS[3]		SS15[3,4]				9034		CS51[3,4] CS41[3,4]		Salt Water		9752[3,4]
34009	2546				SS15[3,4]				9034		CS41[3,4] CS51[3,4]		Salt Water		9752[3,4]
34011															
34015D			L200[3]									511S[3,4]			
34039SS															
3406	2546[3,4]				S14S-3H[2,3] L11S-3H[2,3,4] SS15[3,4]				9034[3,4]		CS41[3] CS51[3,4]		Salt Water[3,4]		9752[3,4]
3406B	2546[4]				S14S-3H[2,3] L11S-3H[2,3,4] SS15[3,4]				9034[4]		CS41[3,4] CS51[3,4]		Salt Water[4]		9752[3,4]
3407	2546[4]		250[2]		SS15[2,3]				9034[4]		C51[3] CS41[3,4]		Salt Water[4]		9752[3,4]
3407SS					Trey Combs Big Game						CS52[3,4] Z8[3,4]				
3408B			178[3,4] ***L064B***[3,4]												
3498AD								B800[3,4] K80[3,4]	8808[3,4]		CS11/1[3,4] CS11/2[3,4] CS11GRS[3,4] ***CS2002***[3,4,5]				9283[3,4]

Hooks that are ***bold italicized*** are barbless.
[1]Mustad considers this hook to be similar. The brand at the top of the column may or may not concur.
[2]The brand at the top of the column considers the Mustad hook to the left to be similar. Mustad may or may not concur.
[3]Similar in all aspects except the bend. See the corresponding characteristic tables.
[4]Similar in all aspects except the finish. See the corresponding characteristic tables.
[5]Similar in all aspects except the point position. See the corresponding characteristic tables.

Mustad	Daiichi	Dai-Riki	Eagle Claw	Gaelic Supreme	Gamakatsu	Hayabusa	Heritage	Kamasan	Orvis	Owner	Partridge	TMC	Turrall	Varivas	VMC
3551											X3 BL[3,4] X3 SI[3,4] X1 BR[3,5] X1 SI[3,4,5] X1 BL[3,4,5] GRS X8[3,4] CS9 BL[3,4,5] CS9 SI[3,4,5] GRS X1[3,4,5]		Tube Fly Treble[3,4]		
35647								B990			X3 BL[3,4] X3 SI[3,4]		Tube Fly Treble[4]		
3582 Discontinued															
3582C Discontinued	7131[3]			Double Wet[3]				B280[3]	0528[3]		P[3]		Salmon Double[3]		
3582F Discontinued	7131[3]			Double Wet[3]				B280[3]	0528[3]		O2		Salmon Double[3]		
36620															
3665A	2340[2,3]								1511		D7A[3]				
36717					Russian River[2,3]										
36890	2441[2,3] 2421[3]				T10-6H[2]	771[3]		B190[2,3] B180[3]	1645[2,3]		M[3]	7999[3]	Salmon Single		
37140															
37141															
37142															
37160			L144[2,3]	Caddis[3,4]	C12U[2,3]	*388*[3,5]						***205BL***[2,5]			
37160S															
37161			L144[3,4]	Caddis[3,4]	C12U[3,4]	*388*[3,4,5]						***205BL***[3,4,5]			
37162			L144[3,4]	Caddis[3,4]	C12U[3,4]	*388*[3,4,5]						***205BL***[3,4,5]			

Mustad	Daiichi	Dai-Riki	Eagle Claw	Gaelic Supreme	Gamakatsu	Hayabusa	Heritage	Kamasan	Orvis	Owner	Partridge	TMC	Turrall	Varivas	VMC
37187	2720 2722[4]														
38941		710[3]	L058[3]						1526[3]		D4A[3] SH3[4]		Extra-long Streamer[3] ***Barbless Extra-long***[3] Bead Streamer[3]		
3906	1550[2]	075[2]	L057[2]		L10-2H[2] S10[3,4]	373		B175[2]	1641[2]		G3A J1A[3] CS7MW[3,4]		Sproat ***Barbless Sproat*** Bead Standard Nymph[3]	98r[3,4]	8526
3906B	1560[2]	060[2] 300[3]						B170	167T[2]		SH2[3,4]		Extra-long Nymph[3] Bead Extra-long Nymph[3]		8527
3908 Discontinued	1530[2]								167T		G3A				
77660SS			L054SS[2]								Z15[3,4]				
7948A	1550[3]	305	L057[3]			373[3]					G3A[3] J1A[3] CS7MW[3,4]		Sproat[3] ***Barbless Sproat***[3] Bead Standard Nymph[3]	98r[3,4]	8526[3]
7957B		300			S12S-1F[3]			B170[3]			L2A[3]		Extra-long Nymph Bead Extra-long Nymph[3]		8527[3]
79580	2220[2]	700[2]	L281[2]		S11-4L2H[2,3]			K80 B800[3]	8808[2]		D4A GRS4A[4] CS11/1[3,4] CS11/2[3,4] CS11GRS[3,4] ***CS2002***[5]	300[3]			9283
79666S															
7970			L1197B[3] L1197G[3,4] L1197N[3,4] NT1197[3,4]												

Hooks that are ***bold italicized*** are barbless.
[1]Mustad considers this hook to be similar. The brand at the top of the column may or may not concur.
[2]The brand at the top of the column considers the Mustad hook to the left to be similar. Mustad may or may not concur.
[3]Similar in all aspects except the bend. See the corresponding characteristic tables.
[4]Similar in all aspects except the finish. See the corresponding characteristic tables.
[5]Similar in all aspects except the point position. See the corresponding characteristic tables.

Mustad	Daiichi	Dai-Riki	Eagle Claw	Gaelic Supreme	Gamakatsu	Hayabusa	Heritage	Kamasan	Orvis	Owner	Partridge	TMC	Turrall	Varivas	VMC
80000BR	***1190[2]*** 1100 1180	305	L059 L060[3] ***L061B***		S10-3F[3]	572[3]		B400[3]	1523		L3A[3] ***L3AY[3]*** GRS3A[3,4]	5210 100[3] ***100BL[3]*** 102Y[3,4] 146Z[4] ***103BL[3,4]*** ***900BL[3,4,5]***			V7061 9280[3]
80010BR	***1190[2]*** 1100 1180	305	L059 L060[3] ***L061B***		S10-3F[3]	572[3]		B400[3]	1523		L3A[3] ***L3AY[3]*** GRS3A[3,4]	5210 100[3] ***100BL[3]*** 102Y[3,4] 146Z[4] ***103BL[3,4]*** ***900BL[3,4,5]***			V7061 9280[3]
80050BR	1270 1273[4]	270[2]	L052[2,3]					B220[3]	1510[2,3]		K12ST GRS12ST[3]				
80060BR	1270 1273[4]							B220[3]			K12ST GRS12ST[3]				
80100BR	1140[2]														
80150BR	1770					380			1512		K6ST[3]	400T[3]			
80200BR	1150[2,3]	135[2,3]	L056[2,3]			387[3]	Grub[3]	B110[3]	8891[2,3]			2457[3]			
80250BR	1130[2,3]	135			C12[2,3]			B100[3] B100G[3,4] B100N[3,4]	1639[2,3]		K4A	2487[3] ***2487BL[3]*** 2487G[3,4]			
80260BR	1130[3]	135			C12[3]			B100[3] B100G[3,4] B100N[3,4]			K4A	2487[3] ***2487BL[3]*** 2487G[3,4]			
80300BR	2720[2,3]								8810[2,3]						
80400BLN	2462[3] 2461[3,4] 2460[3,4]							B940[3,4]		5220[3]					

Mustad	Daiichi	Dai-Riki	Eagle Claw	Gaelic Supreme	Gamakatsu	Hayabusa	Heritage	Kamasan	Orvis	Owner	Partridge	TMC	Turrall	Varivas	VMC
80400D	2462[3,4] 2461[3,4] 2460[3,4]							B940[3,4]		5220[3,4]					
80500BL				F-F Low Water[3]			Single Salmon Low Water[3]		1645[2,3]		O1[3] CS2[3] CS11/3[3,4]		Salmon Low-water		
80501BL											CS2[3] CS11/3[3,4]				
80525BL	7131[2]								0528[2,3]				Salmon Double Low-water		
80526BL															
80550BL															
81000BR	1140[3]								1639[3]						
81001BR						371					K2B		Sedge		
81002BR[6]	1730	700-B			C11-5L2H			B810							
90233S			L200[3,4]									511S[3,4]			
90240	2421[2,3]			F-F Light Wire[3]	T10-3H[2]				1644[2,3]				Salmon Extra-fine Dry Fly		
9049	2421[3]								1644[3]			7989[3]			
9174	X510[3]		NT156[3] L256[2]												
9175	X510[3,4]														
94720															
9479					C14S[2,3,4]										
94831	1280[2]	730			P10-2L1H[2]	552			1638[2]		H1A[3]	5212			V7062
94833		300[2]			S10-3F[2,4]						L4A	5230 5210	Midge		9288

Hooks that are ***bold italicized*** are barbless.
[1]Mustad considers this hook to be similar. The brand at the top of the column may or may not concur.
[2]The brand at the top of the column considers the Mustad hook to the left to be similar. Mustad may or may not concur.
[3]Similar in all aspects except the bend. See the corresponding characteristic tables.
[4]Similar in all aspects except the finish. See the corresponding characteristic tables.
[5]Similar in all aspects except the point position. See the corresponding characteristic tables.

Mustad	Daiichi	Dai-Riki	Eagle Claw	Gaelic Supreme	Gamakatsu	Hayabusa	Heritage	Kamasan	Orvis	Owner	Partridge	TMC	Turrall	Varivas	VMC
94834 Discontinued											K1A[4]				
94838	1310[2,3]				S10-2S[2,3,4]	752[3] 751[3]		B405			E6A[3] ***CS20***[3,4,5]			981[3,4]	9288 9282[3] ***9388***[3]
94840	1170[2] 1100	305[2]	L059[2] ***L061B*** L060[3]		S10[2,4] P-10[2] S10-3F[4]	572[3]		B401[2,3] B402 B400[3]	1523[2]		L2A[3] GRS2A[3,4] L3A[3] ***L3AY***[3] GRS3A[3,4]	5210 100[3] ***100BL***[3] 102Y[3,4] ***103BL***[3,4] ***900BL***[3,4,5] 146Z[4]			V7061 9288 9280[3]
94842	1330[2]				S10U[2,3]			B440[2] K18[3]							9281[3]
94845	***1190***[2] 1100	305	***L061B***[2] L059 L060[3]		***S10-B***[2] S10[4] S10-3F[4]	572[3]		B402[2] B401[3] B400[3]	***1877***[2,5] 1523		L2A[3] GRS2A[3,4] L3A[3] ***L3AY***[3] GRS3A[3,4]	5210 100[3] ***100BL***[3] 102Y[3,4] ***103BL***[3,4] ***900BL***[3,4,5] 146Z[4]			V7061 9288 9280[3]
94859	1480[2,3] 1110	310[2]			S10S[2,4] S-13S-M[2,4] S12S-1F[3]			B410[2,3]	170T[3,4]			101		988[3,4]	
94863 **Discontinued**				F-F Light Wire[3] Single Dry Fly[3]									Salmon Extra-fine Dry Fly[3,4]		
9523															
9575											CS17[4] CS2[3,4] CS11/3[3,4]				
9671		730[2]	L063[2]		L10-2H[2,3]			B830	1524[2] 1638[3]		D4A	5262			
9672	1720[2]	710[2]	L058[2]						1526[2]		SH3[3,4]	5263	Extra-long Streamer ***Barbless Extra-long*** Bead Streamer[3]		

Mustad	Daiichi	Dai-Riki	Eagle Claw	Gaelic Supreme	Gamakatsu	Hayabusa	Heritage	Kamasan	Orvis	Owner	Partridge	TMC	Turrall	Varivas	VMC
9674	1750[2]				S11S-4L2H[2,3]			B820	0167						
9674A	1750[2,4]							B820[4]	0167[4]						
9674B	1750[2,4]							B420[4]	0167[4]						
AC3906B	1560[2]	060[2] 300[3]						B170			SH2[3,4]		Extra-long Nymph[3] Bead Extra-long Nymph[3]		8527
AC94840	1170[2] 1180[2] 1100	305[2]	L059[2] ***L061B*** L060[3]		S10[4] S10-3F[4]	572[3]		B402 B401[3] B400[3]	1523		L2A[3] GRS2A[3,4] L3A[3] ***L3AY***[3] GRS3A[3,4]	5210 100[3] ***100BL***[3] 102Y[3,4] ***103BL***[3,4] ***900BL***[3,4,5] 146Z[4]			V7061 9288 9280[3]
AC9671		730[2]	L063		L10-2H[2,3]			B830	1524 1638[3]		D4A	5262			
AC9672	1720[2]	710[2]	L058[2]						1526[2]		SH3[3,4]	5263	Extra-long Streamer ***Barbless Extra-long*** ***Bead Streamer***[3]		

Hooks that are ***bold italicized*** are barbless.
[1]Mustad considers this hook to be similar. The brand at the top of the column may or may not concur.
[2]The brand at the top of the column considers the Mustad hook to the left to be similar. Mustad may or may not concur.
[3]Similar in all aspects except the bend. See the corresponding characteristic tables.
[4]Similar in all aspects except the finish. See the corresponding characteristic tables.
[5]Similar in all aspects except the point position. See the corresponding characteristic tables.

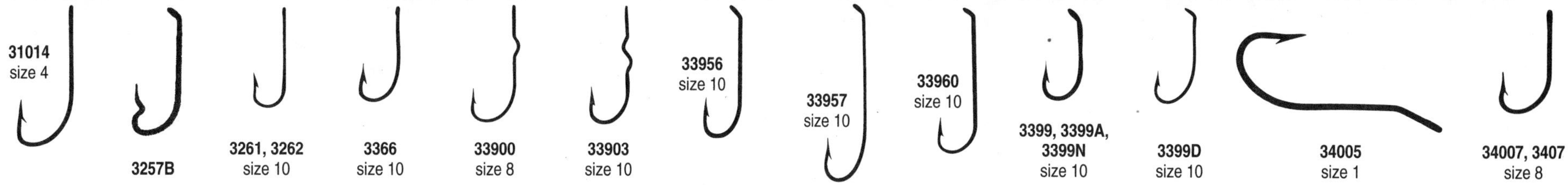

Hook	Use	Wire	Shank	Eye	Bend	Finish	Barb	Point	Remarks	Size
31014	Saltwater	S	S	S	Lim	SS				4/0-4
3257B	Hook Discontinued	S	S	TD	Spr	Bron	Rmks		Kink on spear instead of a barb.	N/A
3261	Bass	S	3XL	S	Aber	Bron				5/0-14
3262	Bass	S	3XL	S	Aber	Blue				5/0-14
3366	Nymph	S	S	S	Spr	Bron			Dry flies, bass.	5/0-14
33900	Popper	S	S-PK	S	Spr	Bron			Single PK.	1/0-12
33903	Bass	S	1XL-PK	S	Spr	Bron			Double PK. One up and one down.	1/0-14
33956	Streamer	S	S+1/4″	TD-U	Spr	Bron			Terrestrials.	8-14
33957	Streamer	S	S+1/2″	TD-U	Spr	Bron			Equal to a 6XL.	2-10
33960	Nymph	S	4XL	TD	Spr	Bron			Wet flies, streamers.	6-10
3399	Nymph	S	S	TD	Spr	Bron			Wet flies. Same as 3399N.	6/0-18
3399A	Nymph	S	S	TD-U	Spr	Bron			Wet flies.	6/0-18
3399D	Dry	2XF	S	TD-U	Spr	Bron				8-16
3399N	Nymph	S	S	TD	Spr	Gold			Wet flies. Same as 3399.	10/0-2
34005	Saltwater	2XH	S-BU	S	Osh	SS			Forged.	5/0-1
34007	Saltwater	S	S	S	Osh	SS			Tube flies. Same as 3407. Forged.	11/0-8

MUSTAD Characteristics

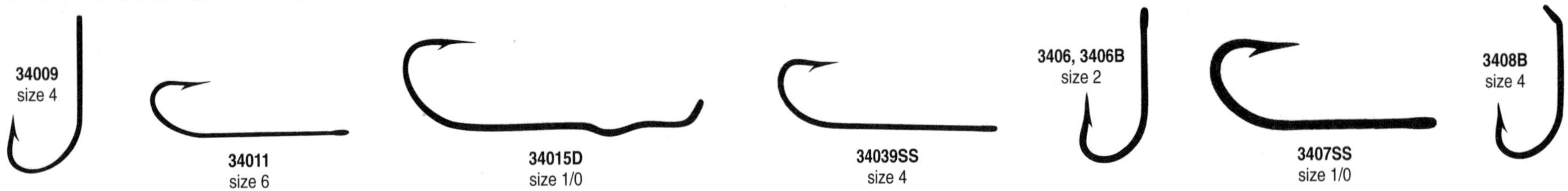

Hook	Use	Wire	Shank	Eye	Bend	Finish	Barb	Point	Remarks	Size
34009	Saltwater	S	S	S-O	Osh	SS			Forged.	5/0-4
34011	Saltwater	S	3XL	S	Osh	SS			Forged.	4/0-6
34015D	Bass	S	2XL-PK	TD	Osh	Corr-Sil			Duratin finish.	1/0
34039SS	Saltwater	S	S+3/8″	S	Osh	SS		Chem	Forged.	2/0-4
3406	Wet	S	S	S	Osh	Bron			Steelhead/salmon. Same as 3406B. Forged.	10/0-2
3406B	Wet	S	S	S	Osh	Nick			Steelhead/salmon. Same as 3406. Forged.	5/0-4
3407	Saltwater	S	S	S	Osh	Tin			Tube flies. Same as 34007. Forged.	14/0-8
3407SS	Saltwater	2XH	S	S	Osh	Tin			Forged.	12/0-1/0
3408B	Saltwater	S	S	TD	Osh	Tin			Forged.	8/0-4

Characteristics MUSTAD

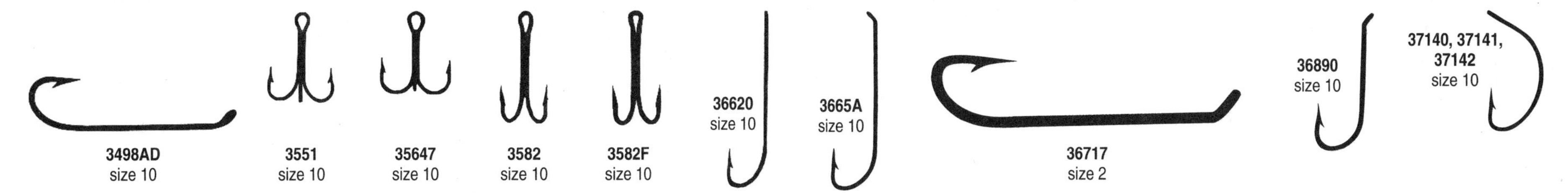

Hook	Use	Wire	Shank	Eye	Bend	Finish	Barb	Point	Remarks	Size
3498AD	Saltwater	S	4XL	TD	Osh	Corr-Sil			Pacific bass. Duratin finish.	5/0-4, 10, 8
3551	Treble	S	S	S	Lim-Treb	Bron			Tube flies.	5/0-20
35647	Treble	S	S	S	P-Treb	Bron			Tube flies.	3/0-14
3582	Hook Discontinued	S	S	TD-L	N.I.-Dbl	Bron			Steelhead/salmon. Hook replaced by 80000 series.	N/A
3582C	Hook Discontinued	S	S	TU-L	N.I.-Dbl	Blk			Oval eye. Hook replaced by 80000 series.	N/A
3582F	Hook Discontinued	S	S	TU-L	N.I.-Dbl	Blk			Steelhead/salmon. Hook replaced by 80000 series.	N/A
36620	Streamer	S	S+1/2″	S	Lim	Bron			Thunder Creek flies. Equal to a 6XL.	4-10
3665A	Streamer	S	S+1/2″	TD	Lim	Bron			Equal to a 6XL.	2-12
36717	Streamer	5XH	S+1/2″	TD	Lim	Bron			Equal to a 6XL. Meets Russian River regulations.	2
36890	Steelhead/Salmon	S	S	TU-L	Lim	Blk			Forged.	3/0-12
37140	Scud/Shrimp	S	S-C	S	C-WG-R	Bron			Same as 37141, 37142.	6/0-14
37141	Scud/Shrimp	S	S-C	S	C-WG-R	Bron			Same as 37140, 37142.	6/0-14
37142	Scud/Shrimp	S	S-C	S	C-WG-R	Bron			Same as 37140, 37141.	6/0-14

Characteristics

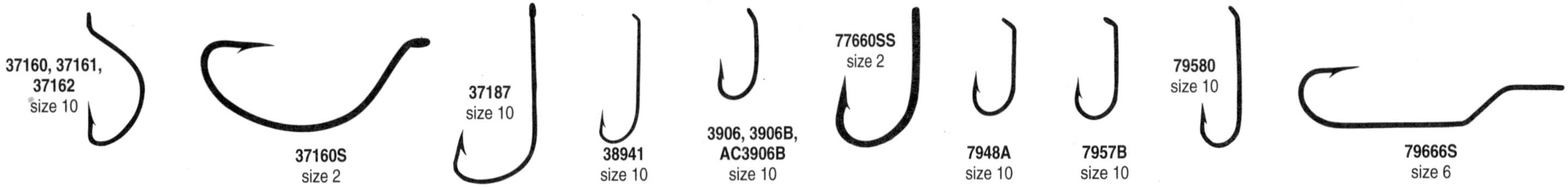

Hook	Use	Wire	Shank	Eye	Bend	Finish	Barb	Point	Remarks	Size
37160	Scud/Shrimp	S	S-C	TU	C-WG-R	Bron			Caddis larva. Same as 37161, 37162.	3/0-20
37160S	Saltwater	1XF	1XS-C	TU	C-WG-R	SS				5/0-2
37161	Scud/Shrimp	S	S-C	TU	C-WG-R	Nick			Caddis larva. Same as 37160, 37162.	6/0-12
37162	Scud/Shrimp	S	S-C	TU	C-WG-R	Gold			Caddis larva. Same as 37160, 37161.	4/0-10
37187	Bass	S	S	S	WG	Bron			Bass bugs. Open bend.	1/0-10
38941	Streamer	S	3XL	TD	Spr	Bron			Nymphs, Muddlers.	2-14
3906	Nymph	S	S	TD	Spr	Bron			Wet flies.	2-20
3906B	Nymph	1XH	1XL	TD	Spr	Bron		Chem	Same as AC3906B.	4-20
3908	Hook Discontinued	1XH	S	TD	Spr	Bron			Steelhead/salmon.	N/A
77660SS	Saltwater	1XH	1XS	S	Osh	SS		Chem	Tarpon, tube flies. Forged.	6/0-2
7948A	Nymph	S	S	TD	P	Bron			Wet flies. Forged.	6-16
7957B	Nymph	S	1XL	TD	P	Bron			Wet flies. Forged.	6-18
79580	Streamer	S	4XL	TD	P	Bron			Forged.	2-16
79666S	Saltwater	S	S-K	S	P	SS			Bass bugs. Forged.	1/0-6

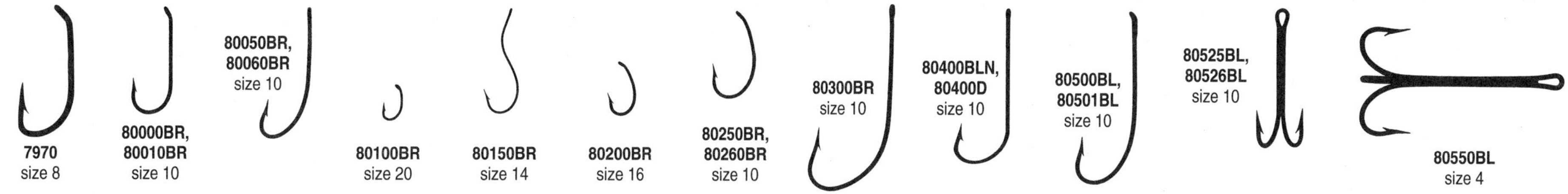

Hook	Use	Wire	Shank	Eye	Bend	Finish	Barb	Point	Remarks	Size
7970	Steelhead/Salmon	5XH	S	TD	Lim	Bron			Bass.	1-8
80000BR	Dry	1XF	S	TD	P	Bron	Mini	Chem	Same as 80010BR. Forged.	2-26
80010BR	Dry	1XF	S	TD	P	Bron	No Barb	Chem	Same as 80000BR. Forged.	10-14
80050BR	Nymph	S	3XL	S	York	Bron	Mini	Chem	Terrestrials, dry flies. Same as 80060BR. Forged.	4-22
80060BR	Nymph	S	3XL	S	York	Bron	No Barb	Chem	Same as 80050BR. Forged.	10-14
80100BR	Midge/Pupae	1XF	1XS-C	TD	C-WG	Bron	Mini	Chem	Tricos. Forged.	20-26
80150BR	Nymph	1XF	3XL-BU	S	Spr	Bron	Mini	Chem	Swimming nymph. Bent up 45°. Forged.	8-14
80200BR	Scud/Shrimp	2XH	1XS-C	TD	C-WG-R	Bron	Mini	Chem	SJW, pupae. Forged.	6-16
80250BR	Scud/Shrimp	S	1XS-C	TD	C-WG-R	Bron	Mini	Chem	SJW, pupae. Same as 80260BR. Forged.	10-22
80260BR	Scud/Shrimp	S	1XS-C	TD	C-WG-R	Bron	No Barb	Chem	SJW, pupae. Same as 80250BR. Forged.	12-16
80300BR	Bass	1XF	2XL	S	WG	Bron	Mini	Chem	Forged.	2-10
80400BLN	Streamer	S	3XL	S	P-M-R	Nick	Mini	Chem	Same as 80400D. Forged.	2-12
80400D	Streamer	S	3XL	S	P-M-R	Corr-Sil	Mini	Chem	Same as 80400BLN. Forged.	2-12
80500BL	Steelhead/Salmon	S	2XL	TU-L	Lim	Blk		Chem	Forged.	3/0-10
80501BL	Steelhead/Salmon	S	2XL	TD-L	Lim	Blk		Chem	Forged.	3/0-10
80525BL	Double	S	2XL	TU-L	Lim-Dbl	Blk		Chem	Steelhead/salmon. Forged.	2-10
80526BL	Double	S	2XL	TD-L	Lim-Dbl	Blk		Chem	Forged.	2-10
80550BL	Treble	S	2XL	TD	Lim-Treb	Blk		Chem	Forged.	1/0-4

MUSTAD   Characteristics

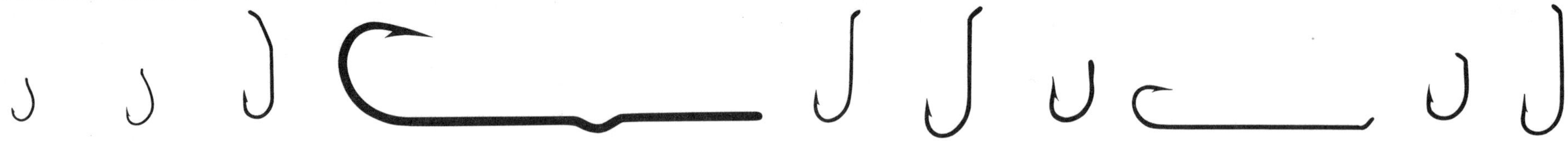

81000BR size 18 · 81001BR size 16 · 81002BR size 14 · 90233S size 2/0 · 90240 size 10 · 9049 size 8 · 9174, 9175 size 10 · 94720 size 8 · 9479 size 10 · 94831 size 10

Hook	Use	Wire	Shank	Eye	Bend	Finish	Barb	Point	Remarks	Size
81000BR	Scud/Shrimp	1XF	1XS-C	TU-M	C-WG-R	Bron		Chem	SJW, pupae. Eye partially TU. Forged.	18-22
81001BR	Scud/Shrimp	1XH	S-C	TU	C-WG	Bron			SJW, pupae. Forged.	8-16
81002BR	Nymph	1XH	4XL-BD	TD	P	Bron			Wet flies. Shank bent down 20°. Forged.	6-14
90233S	Saltwater	S	4XL-PK	S	Spr-WG	SS			Bass, pike.	2/0
90240	Steelhead/Salmon	2XF	1XL	TU-L	Lim	Blk				4-10
9049	Steelhead/Salmon	1XF	1XL	TU-L	Lim	Blk				4-8
9174	Egg	1XH	3XS	S	Osh	Bron			Same as 9175. Forged.	9/0-10
9175	Egg	1XH	3XS	S	Osh	Tin			Same as 9174. Forged.	9/0-2
94720	Streamer	S	8XL	TD	P	Bron			Forged.	2-8
9479	Egg	1XF	5XS	TD	P-R	Bron			Midges. Forged.	2-12
94831	Dry	2XF	2XL	TD	P	Bron			Forged.	6-16

Hook	Use	Wire	Shank	Eye	Bend	Finish	Barb	Point	Remarks	Size
94833	Dry	3XF	S	TD	P	Bron			Forged.	6-20
94834	Hook Discontinued	1XF	S	TD	P	Gold			Forged. Was available in size 28 only.	28
94838	Dry	1XF	1XS	TD	P-M	Bron			Bend is wider than standard but not as wide as WG. Forged.	12-18
94840	Dry	1XF	S	TD	P	Bron			Same as AC94840 & 94845. Forged.	2-28
94842	Dry	1XF	S	TU	P	Bron			Midges, pupae, Tricos. Forged.	8-24
94845	Dry	1XF	S	TD	P	Bron	No Barb		Same as 94840 & AC94840. Forged.	8-20
94859	Dry	1XF	S	S	P	Bron			Midges, spiders. Forged.	18-28
94863	Hook Discontinued	2XF	S	TU-L	P	Bron	No Barb		Forged.	N/A
9523	Dry	1XF	5XS	TU	P-R	Bron			Forged.	2-16
9575	Streamer	S	S+$^1/_2$″	TD-L	Lim	Bron			Equal to a 6XL. Forged.	2-12
9671	Nymph	S	2XL	TD	P	Bron			Same as AC9671. Forged.	2-18
9672	Streamer	S	3XL	TD	P	Bron			Nymphs. Same as AC9672. Forged.	2-18
9674	Streamer	S	4XL	S	P	Bron			Same as 9674A & 9674B. Forged.	1-10
9674A	Streamer	S	4XL	S	P	Gold			Same as 9674 & 9674B. Forged.	8-14
9674B	Streamer	S	4XL	S	P	Nick			Same as 9674 & 9674A. Forged.	2-8
AC3906B	Nymph	1XH	1XL	TD	Spr	Bron		Chem	Same as 3906B.	10-16
AC94840	Dry	1XF	S	TD	P	Bron		Chem	Same as 94840 & 94845. Forged.	8-18
AC9671	Streamer	S	2XL	TD	P	Bron		Chem	Same as 9671. Forged.	8-14
AC9672	Streamer	S	3XL	TD	P	Bron		Chem	Nymphs. Same as 9672. Forged.	4-12

ORVIS Cross-Reference

Orvis	Daiichi	Dai-Riki	Eagle Claw	Gaelic Supreme	Gamakatsu	Hayabusa	Heritage	Kamasan	Mustad	Partridge	TMC	Turrall	Varivas	VMC
Xpoint	X452		250[4]					K93[3,4]						7255[4]
0167	1750[1,2]							B820	9674 9674A[4] 9674B[4]		9395[1,3]			
167T		075	L1197B L1197G[4] L1197N[4] NT1197[4]					B175	3906B[1]		3761[1] 700[3,4,5] 9300[3]			
170T	1110[3,4]				S12S-1F[3,4] S-13S-M[3]				94859[3,4]		101[3,4]			
0528	7131[1,2,3]								80525BL[1,3]					
594F	1740[1,2]													
596F	2370[1,2]													
1509	1310[1,2,3] 1170[3] 1180[3] 1182[3,4] ***1190***[3]				P-10[3] S10[3,4] ***S10-B***[3,4]	372[3] 373[3] ***378***[3] 386[3]		B401[3] ***B402***[3] K40[3]			100[1]		980[3]	9060[3]
1510	1270[1,2,3] 1273[3,4]	270[1,3]				383[3]			80050BR[1]		200R[1,3,5] ***200RBL***[3,5]			
1511	2340[1,2]			MMRS-6XL			Allcock Streamer[3]		3665A	D7A[3]	300[1,3]			9283[1,3]
1512	1770[1,2]					380			80150BR	K6ST[3]	400T[1,3]			
1523	1180[1,2] 1100		L059 L060[3] ***L061B***		S10-3F[4]	572[3]		B400[3]	94840[1] AC94840 ***94845*** 80000BR ***80010BR***	L3A[3] ***L3AY***[3] GRS3A[3,4]	5210[1] 100[3] ***100BL***[3] 102Y[3,4] ***103BL***[3,4] ***900BL***[3,4,5] 146Z[4]			9280[1,3] V7061
1524	1710[1,2]	730[1]	L063					B830	9671[1] AC9761	H1A[3]	5262[1]			9279[1,3]

Hooks that are ***bold italicized*** are barbless.
[1]Orvis considers this hook to be similar. The brand at the top of the column may or may not concur.
[2]The brand at the top of the column considers the Orvis hook to the left to be similar. Orvis may or may not concur.
[3]Similar in all aspects except the bend. See the corresponding characteristic tables.
[4]Similar in all aspects except the finish. See the corresponding characteristic tables.
[5]Similar in all aspects except the point position. See the corresponding characteristic tables.

Orvis	Daiichi	Dai-Riki	Eagle Claw	Gaelic Supreme	Gamakatsu	Hayabusa	Heritage	Kamasan	Mustad	Partridge	TMC	Turrall	Varivas	VMC
1526	1720[1,2]	710							9672[1] AC9672 38941[3]	SH3[3,4]	5263[1]	Extra-long Streamer ***Barbless Extra-long*** Bead Streamer[3]		
1638	1280[2]		L063[3]					B830[3]	94831[1] 9671[3] AC9671[3]		5212[1]			
1639	1140[1,2] 1130[1,2]	135[1]							80250BR[1]		2487[1]			
1641	1550[2] 1530	075[1]							3906[1]		3769[1]			8526[1]
1644				Single Dry Fly[3]					90240[1,3] 9049[3]		7989[1,3]			
1645	2441[2] 2421[2,3]			Single Low Water[3]				B190[3]	36890[1,3] 80500BL[1,3]	M[3]	7999[1,3]		974[3]	
1877	***1190***[2,5] 1170[5] 1180[5] 1182[4,5]	305[5]			P-10[5] S10[4,5] ***S10-B***[4,5]	372[5] 373[3,5] ***378***[5] 386[5]		B401[5] ***B402***[5] K40[5]	***94845***[1,5]		***103BL***[1,3,4]		980[5]	9060[5]
4641	1110[2,3]	310[1,3]												
4864	1100[2] 1170[3] 1180[3] 1182[3,4] ***1190***[3]	305[3]			P-10[3,5] S10[3,4] ***S10-B***[3,4]	372[3] 373[3] ***378***[3] 377[3] ***379***[3] 386[3]		B401[3] ***B402***[3] K40[3]					980[3]	9060[3]
8808	2220[1,2]	700						B800[3] K80	79580[1] 33960[3] 3498AD[3,4]	CS11/1[3,4] CS11/2[3,4] CS11GRS[3,4] ***CS2002***[5]				9283
8810	2720 2722[4]								80300BR[1]		8089[1,3] 8089NP[3,4]			

Orvis	Daiichi	Dai-Riki	Eagle Claw	Gaelic Supreme	Gamakatsu	Hayabusa	Heritage	Kamasan	Mustad	Partridge	TMC	Turrall	Varivas	VMC
8891	1150[3] 1155[3,4]								80200BR[1,3]	K2B				
9034	2546	930			SS15[3,4]				34007 34009 31014[3] 3407[4] 3366[3,4] 3406B[4] 3406[4]	CS51[3,4] CS41[3,4]		Salt Water		9752[3,4]

Hooks that are ***bold italicized*** are barbless.
[1]Orvis considers this hook to be similar. The brand at the top of the column may or may not concur.
[2]The brand at the top of the column considers the Orvis hook to the left to be similar. Orvis may or may not concur.
[3]Similar in all aspects except the bend. See the corresponding characteristic tables.
[4]Similar in all aspects except the finish. See the corresponding characteristic tables.
[5]Similar in all aspects except the point position. See the corresponding characteristic tables.

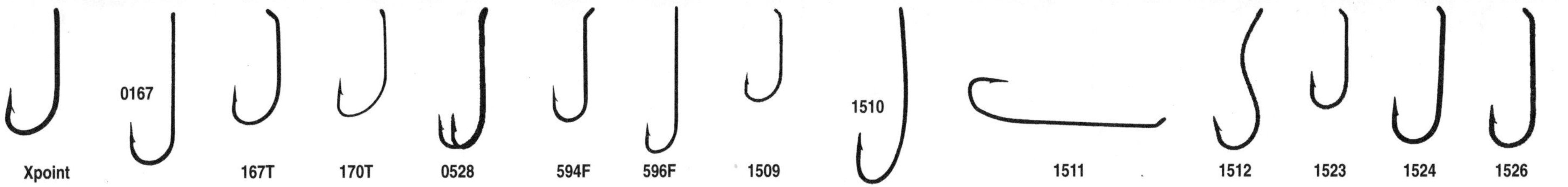

Hook	Use	Wire	Shank	Eye	Bend	Finish	Barb	Point	Common Name	Remarks	Size
XPoint	Saltwater	S	1XL	S	Osh	Sil		Chem	XPoint	Salmon trailer, bass, tube flies. Forged.	4/0-8
0167	Streamer	S	4XL	S	P	Bron	Mini	Chem	Straight-Eye Streamer	Bucktails, leeches, Woolly Buggers. Forged.	4-14
167T	Wet	1XH	S	TD	Spr	Bron	Mini	Chem	Standard Wet Fly		6-18
170T	Dry	1XF	S	S	Lim	Blk	Mini	Chem	Limerick Dry Fly		12-24
0528	Double	1XH	S	TU-L	Spr-Dbl	Blk	Mini	Chem	Double Salmon		4-12
594F	Nymph	1XH	2XL	TU	P	Bron	Mini	Chem	Up-Eye Nymph		6-16
596F	Streamer	1XH	7XL	S-L	Lim-M	Bron	Mini	Chem	Loop-Eye Streamer	Display flies.	2-10
1509	Dry	S	S	TD	P-M-WG	Bron	Mini	Chem	Wide Gape Dry Fly		10-24
1510	Nymph	S	3XL-C	S	N.I.-M	Bron	Mini	Chem	Curved Nymph		6-22
1511	Streamer	S	6XL	TD	Lim	Bron	Mini	Chem	Streamer		4-12
1512	Nymph	1XF	3XL-BU	S	Spr	Bron	Mini	Chem	Swimming Nymph	Swimming nymphs.	6-16
1523	Dry	1XF	S	TD	P	Bron	Mini	Chem	Extra-Fine Dry Fly		8-22
1524	Nymph	S	2XL	TD	P	Bron	Mini	Chem	Traditional Nymph		4-18
1526	Nymph	S	3XL	TD	P	Bron	Mini	Chem	Muddler/Stonefly	Muddlers, sculpins, Zonkers, stoneflies.	6-18

Characteristics

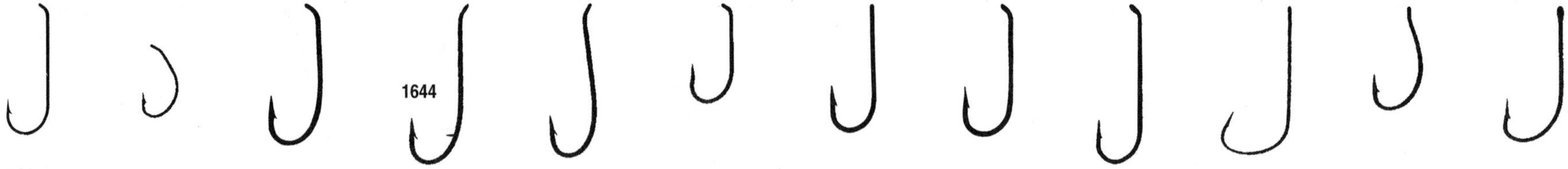

Hook	Use	Wire	Shank	Eye	Bend	Finish	Barb	Point	Common Name	Remarks	Size
1638	Dry	S	2XL	TD	P-M	Bron	Mini	Chem	2X Dry Fly	Terrestrials, long-bodied dry flies. Smaller gape.	6-16
1639	Scud/Shrimp	1XF	1XS-C	Rmks	C-WG	Bron	Mini	Chem	Shrimp/Caddis Nymph	Sizes 10-16 are TD eye; sizes 18-22 are TU eye.	10-22
1641	Wet	2XH	1XS	TD	Spr	Bron	Mini	Chem	Heavy Wet/Nymph	Small steelhead/salmon flies.	2-16
1644	Steelhead/Salmon	1XF	S	TU-L	Spr	Blk	Mini	Chem	Salmon Dry Fly		2-12
1645	Steelhead/Salmon	1XH	S	TU-L	Spr	Blk	Mini	Chem	Salmon Wet Fly		1-8
1877	Dry	S	S	TD	P	Bron	No Barb	SDP-C	Barbless Dry Fly		8-24
4641	Dry	S	S	S-O	P-M	Bron	Mini	Chem	Straight-Eye Big Eye	Wet flies, nymphs, small streamers.	16-24
4864	Dry	S	S	TD-O	P-M	Bron	Mini	Chem	Down-Eye Big Eye	Parachutes.	16-24
8808	Streamer	S	4XL	TD	P	Bron	Mini	Chem	4X Streamer	Nymphs, Woolly Buggers.	2-14
8810	Bass	1XF	S	S	WG	Bron	Mini	Chem	Bass Bug		2, 1/0, 3/0
8891	Caddis	1XH	1XS	TU	C-WG	Bron	Mini	Chem	Heavy Wire Worm/Caddis	SJW	8-16
9034	Saltwater	S	S	S	Osh	SS	Mini		PreSharpened Saltwater		1-6, 4/0-1/0

Owner	Daiichi	Dai-Riki	Eagle Claw	Kamasan	Mustad	TMC
5106	2451[3,4]					
5111						
5170		930[3,4]	L7254[3,4] L054SS[3,4]			811S[3,4] 800S[3,4,5]
5320	2462[3] 2460[3,4] 2461[3,4]			B940[3,4]	3262[3,4] 3261[3,4] 80400BLN[3] 80400D[3,4]	

Hooks that are ***bold italicized*** are barbless.
[1]Owner considers this hook to be similar. The brand at the top of the column may or may not concur.
[2]The brand at the top of the column considers the Owner hook to the left to be similar. Owner may or may not concur.
[3]Similar in all aspects except the bend. See the corresponding characteristic tables.
[4]Similar in all aspects except the finish. See the corresponding characteristic tables.
[5]Similar in all aspects except the point position. See the corresponding characteristic tables.

OWNER Characteristics

5106

5111

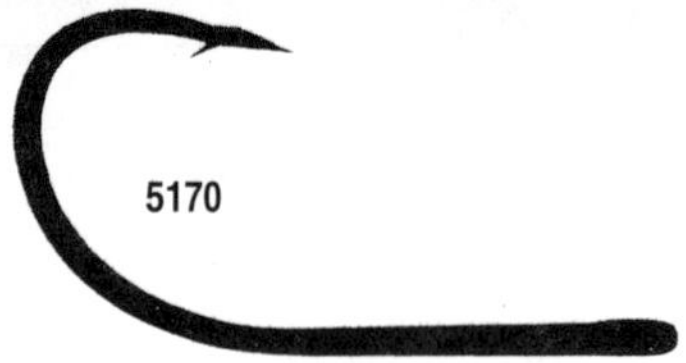

5170

5320

Hook	Use	Wire	Shank	Eye	Bend	Finish	Barb	Point	Common Name	Remarks	Size
5106	Saltwater	S	1XS	S	P-M	Corr-Nick			Flyliner	Forged.	1/0-8
5111	Saltwater	S	1XS	TU	P-M-R	(Rmks)		SDP	SSW with Cutting Point	Finish available in corr-nick (sizes 8/0-8) and corr-red (sizes 4/0-8). Forged.	8/0-8
5170	Saltwater	1XH	S	S	Spr-M	Corr-Nick			AKI Hook	Saltwater streamers. Forged.	6/0-1/0
5320	Saltwater	S	3XL	S	P-WG	Corr-Nick				Saltwater streamers. Forged.	5/0-1/0

Partridge	Daiichi	Dai-Riki	Eagle Claw	Gaelic Supreme	Gamakatsu	Hayabusa	Heritage	Kamasan	Mustad	Orvis	TMC	Turrall	Varivas	VMC
A	1310[3]													
B Discontinued	1330[3]					375[3]		B440[3]						
CS2									80501BL[3] 80500BL[3] 9575[3,4]					
CS5														
CS6				F-F Light Blind Eye			Pryce-Tannant					Salmon Blind Eye		
CS7MW	1550[3,4]	305[3,4]	L057[3,4]		S10[3,4]	373[3]			3906[3,4] 3399[3,4] 3399A[3,4] 3399N[3,4] 7948A[3,4]			Sproat[3,4] ***Barbless Sproat***[3,4] Bead Standard Nymph[4]	98r[3]	8526[3,4]
CS8ST														
CS9 BL									3551[3,4,5]					
CS9 SI									3551[3,4,5]					
CS10/1														
CS10/2														
CS10/3	2091[3]													
CS10/4														
CS11/1								B800[3,4] K80[3,4]	3498AD[3,4] 33960[3,4] 79580[3,4]	8808[3,4]				9283[3,4]
CS11/2								B800[3,4] K80[3,4]	3498AD[3,4] 33960[3,4] 79580[3,4]	8808[3,4]				9283[3,4]
CS11/3									9575[3,4] 80500BL[3,4] 80501BL[3,4]					

Hooks that are ***bold italicized*** are barbless.
[1]Partridge considers this hook to be similar. The brand at the top of the column may or may not concur.
[2]The brand at the top of the column considers the Partridge hook to the left to be similar. Partridge may or may not concur.
[3]Similar in all aspects except the bend. See the corresponding characteristic tables.
[4]Similar in all aspects except the finish. See the corresponding characteristic tables.
[5]Similar in all aspects except the point position. See the corresponding characteristic tables.

Partridge	Daiichi	Dai-Riki	Eagle Claw	Gaelic Supreme	Gamakatsu	Hayabusa	Heritage	Kamasan	Mustad	Orvis	TMC	Turrall	Varivas	VMC
CS11GRS								B800[3,4] K80[3,4]	3498AD[3,4] 33960[3,4] 79580[3,4]	8808[3,4]				9283[3,4]
CS12				Treble Low Water[3] Treble Wet[3]				B380[3]				Salmon Treble[3,5] Salmon Treble Low-water[3,5]		
CS15				MMRS-10XL[3]					9575					
CS17	2271			Salmon Streamer[3,4]					9575[3]					
CS18														
CS20						751[3]		B405[3,4,5]	94838[3,4,5]		921		981[3]	9282[3,4,5] ***9388***[3,4]
CS21							Trout Xtra Fine[3,4]					***Barbless Extra-fine***[3,4] Extra-fine Dry Fly[3,4]		
CS2001							Trout Xtra Fine[5]					***Barbless Extra-fine***[5] Extra-fine Dry Fly[5]		
CS2002								B800[3,5] K80[5]	33960[3,5] 3498AD[3,4,5] 79580[5]	8808[5]				9283[5]
CS2003														
CS27GRS									3399D[3,4]		***902BL***[3,4]		98i[3,4]	V7060[3,4]
CS28GRS														
CS29GRS														

Hooks that are ***bold italicized*** are barbless.
[1]Partridge considers this hook to be similar. The brand at the top of the column may or may not concur.
[2]The brand at the top of the column considers the Partridge hook to the left to be similar. Partridge may or may not concur.
[3]Similar in all aspects except the bend. See the corresponding characteristic tables.
[4]Similar in all aspects except the finish. See the corresponding characteristic tables.
[5]Similar in all aspects except the point position. See the corresponding characteristic tables.

Partridge	Daiichi	Dai-Riki	Eagle Claw	Gaelic Supreme	Gamakatsu	Hayabusa	Heritage	Kamasan	Mustad	Orvis	TMC	Turrall	Varivas	VMC
CS41	2546[3,4]				SC15[3,4]				3366[3] 31014[3,4] 3407[3,4] 34007[3,4] 3406[3] 3406B[3,4] 34009[3,4]	9034[3,4]		Salt Water[3,4]		9148[4] 9752[3,4]
CS42														
CS51	2546[3,4]				SS15				3366[3,4] 31014[3,4] 3407[3] 34007[3,4] 3406[3,4] 3406B[3,4] 34009[3,4]	9034[3,4]		Salt Water[3,4]		9752[3,4]
CS52			250[3,4]		Trey Combs Big Game[3,4]				3407SS[3,4] 34009[4]		811S[4]			
CS53														
CS54														
D3ST	1750[2,3]							B940[3,4]						9255[3]
D4A	2220[2] 1710[2]	700	L281 L058						79580 9671					9279[3]
D5B														
D7A				MMRS-6XL[3]			Allcock Streamer[3]		3665A[3]	1511[3]				
E1A												Mayfly[3]		
E3AY	***1190***[2,3]						Trout Xtra Fine[3]					***Barbless Extra-fine***[3] Extra-fine Dry Fly[3]		
E4A							Trout Xtra Fine[3]					***Barbless Extra-fine***[3] Extra-fine Dry Fly[3]		
E6A						751		B405[3]	94838[3]		921			9282[3]

Partridge	Daiichi	Dai-Riki	Eagle Claw	Gaelic Supreme	Gamakatsu	Hayabusa	Heritage	Kamasan	Mustad	Orvis	TMC	Turrall	Varivas	VMC
G3A	1550[2]		L057		S10[3,4] ***S10-B***[3,4]	373			3906 3399 3399A 3399N[4] 7948A			Sproat ***Barbless Sproat*** Bead Standard Nymph[3]	98r	8526 8527
GRS 1Z	3111[3,4]		L254N[3,4] 254N[3,4] 254[3,4] L253[3,4] L254BK[3,4] L254SS[3,4] L067[3,4]		SC15[3,4]									7250NI[3,4] 8410[3,4]
GRS2A			L059[3,4] ***L061B***[3,4]						3399D[3,4] 94840[3,4] AC94840[3,4] ***94845***[3,4]		***902BL***[3,4]		98i[3,4]	V7060[3,4] 9280[3,4]
GRS3A	1100[3,4]		L059[3,4] ***L061B***[3,4] L060[3,4]		S10-3F[3,4]	572[3,4]	Trout Xtra Fine[3,4]	B400[3,4]	94833[3,4] 80000BR[3,4] ***80010BR***[3,4] 94840[3,4] AC94840[3,4] ***94845***[3,4]	1523[3,4]	100[3,4] ***100BL***[3,4] 102Y[3,4] ***103BL***[3,4] 146Z[3,4] 5210[3,4] ***900BL***[3,4,5]	***Barbless Extra-fine***[3,4] Extra-fine Dry Fly[3,4]		V7061[3,4] 9280[3,4]
GRS4A	2220[4] 1710[4]	700[4]	L281[4] L058[4]						79580[4] 9671[4]					9279[3]
GRS7MMB														
GRS 13Z														
GRS12ST	1270[3] 1273[3,4]							B220[3,4]	80050BR[3,4] ***80060BR***[3,4]					
GRSQ2				Double Low Water[3,4] Double Wet[3,4]				B280[3,4]				Salmon Double Low-water[3] Salmon Double[3,4] Sproat Double[3,4]		9410[3,4]
GRS S3														

Hooks that are ***bold italicized*** are barbless.
[1]Partridge considers this hook to be similar. The brand at the top of the column may or may not concur.
[2]The brand at the top of the column considers the Partridge hook to the left to be similar. Partridge may or may not concur.
[3]Similar in all aspects except the bend. See the corresponding characteristic tables.
[4]Similar in all aspects except the finish. See the corresponding characteristic tables.
[5]Similar in all aspects except the point position. See the corresponding characteristic tables.

Partridge	Daiichi	Dai-Riki	Eagle Claw	Gaelic Supreme	Gamakatsu	Hayabusa	Heritage	Kamasan	Mustad	Orvis	TMC	Turrall	Varivas	VMC
GRS S4			250H[3,4]		Trey Combs Big Game[3,4]									
GRS X1									3551[3,4,5]					
GRS X2B				Treble Low Water[3,4,5] Treble Wet[3,4,5]				B380[3,4,5]				Salmon Treble[3,4] Salmon Treble Low-water[3,4]		
GRS X8									3551[3,4]					
H1A	1280[2,3]	730[3]	L063[3]			552[3]			94831[3]	1524[3]	5212[3]			V7062[3]
H1B/CA														
H3ST														
J1A	1550[3]		L057[3]		S10[3,4]	373[3]			3906[3] 3399[3] 3399A[3] 3399N[3,4] 7948A[3]			Sproat[3] ***Barbless Sproat***[3] Bead Standard Nymph[3]	98r[4]	8526[3]
K1A	1480[2,3]								98834[4]					
K2B	1150[3] 1155[3,4]					371			81001BR	8891[3]		Sedge		
K3A														
K4A	1130[2]							B100[3] B100G[3,4] B100N[3,4]	80250BR ***80260BR***					
K6ST	1770[3]					380[3]			80150BR[3]	1512[3]	400T[4]			
K7														
K10														
K12ST	1270[3,4] 1273[3,4]							B220[3,4]	80050BR[3] ***80060BR***[3]					
K14ST														

Partridge	Daiichi	Dai-Riki	Eagle Claw	Gaelic Supreme	Gamakatsu	Hayabusa	Heritage	Kamasan	Mustad	Orvis	TMC	Turrall	Varivas	VMC
K16B														
K17														
L2A	1170[2]		L059[3] ***L061B***[3]						3399D[3] 94840[3] AC94840[3] ***94845***[3]		***902BL***[3,4]		98i[3,4]	V7060[3] 9280[3]
L3A	1100[3]		L059[3] ***L061B***[3] L060[3]		S10-3F[3,4]	572[3]	Trout Xtra Fine[3]	B400[3]	94833[3] 80000BR[3] ***80010BR***[3] 94840[3] AC94840[3] ***94845***[3]	1523[3]	100[3] ***100BL***[3] 102Y[3,4] ***103BL***[3,4] 146Z[3,4] ***900BL***[3,4,5] 5210[3]	***Barbless Extra-fine***[3] Extra-fine Dry Fly[3]		9280[3] V7061[3]
L3AY	1100[3]		L059[3] ***L061B***[3] L060[3]		S10-3F[3,4]	572[3]	Trout Xtra Fine[3]	B400[3]	94833[3] 80000BR[3] ***80010BR***[3] 94840[3] AC94840[3] ***94845***[3]	1523[3]	100[3] ***100BL***[3] 102Y[3,4] ***103BL***[3,4] 146Z[3,4] ***900BL***[3,4,5] 5210[3]	***Barbless Extra-fine***[3] Extra-fine Dry Fly[3]		9280[3] V7061[3]
L3B				Dry Fly[3,4]										
L4A									94833		5230[3]			
M	2441[2,3]			F-F Wet Fly[3] Single Wet[3,4] Single Low Water[3]				B190[3]	36890[3]	1645[3]	7999[3]		974[3]	
MM1B														
MM2A	1310[3]													
MM3ST	3111[3]		L254N[3,4] 254N[3,4] 254[3,4] L253[3,4] L254BK[3] L254SS[3,4] L067[3,4]		SC15[3,4]									7250NI[3,4] 8410[3,4]

Hooks that are ***bold italicized*** are barbless.

[1]Partridge considers this hook to be similar. The brand at the top of the column may or may not concur.

[2]The brand at the top of the column considers the Partridge hook to the left to be similar. Partridge may or may not concur.

[3]Similar in all aspects except the bend. See the corresponding characteristic tables.

[4]Similar in all aspects except the finish. See the corresponding characteristic tables.

[5]Similar in all aspects except the point position. See the corresponding characteristic tables.

Partridge	Daiichi	Dai-Riki	Eagle Claw	Gaelic Supreme	Gamakatsu	Hayabusa	Heritage	Kamasan	Mustad	Orvis	TMC	Turrall	Varivas	VMC
N	2421[2,3]			Single Low Water[3]										
O1									80500BL[3]					
O2														
P				Double Wet[3]										
Q1	7131[2]			Double Low Water[3] Double Wet[3]				B280[3]				Salmon Double Low-water[3] Salmon Double[3] Sproat Double[3]		9410[3]
R1A								B270[3]						
R2A								B270[3]				Sproat Double[3]		
R3HF														
SH1		075[3,4]			L10-2H[3,4]						3769[3,4]		977[4]	
SH2	1560[3,4]	060[3,4]							3906B[3,4] AC3906B[3,4]					
SH3		710[3,4]							9672[3,4] AC9672[3,4] 38941[4]	1526[3,4]		Extra-long Streamer[3,4] ***Barbless Extra-long***[3,4] Bead Streamer[3,4]		
S3														
S4			250H[3,4]		Trey Combs Big Game[3,4]									
X1 BL									3551[3,4,5]					
X1 BR									3551[3,5]					
X1 SI									3551[3,4,5]					
X2B				Treble Low Water[3,5] Treble Wet[3,5]				B380[3,5]				Salmon Treble[3] Salmon Treble Low-water[3]		

Partridge	Daiichi	Dai-Riki	Eagle Claw	Gaelic Supreme	Gamakatsu	Hayabusa	Heritage	Kamasan	Mustad	Orvis	TMC	Turrall	Varivas	VMC
X3 BL									3551[3,4] 35647[3,4]			Tube Fly Treble[3]		
X3 SI									3551[3,4] 35647[3,4]			Tube Fly Treble[3,4]		
X7														
Z1	3111[3]		L254N[3,4] 254N[3,4] 254[3,4] L253[3,4] L254BK[3] L254SS[3,4] L067[3,4]		SC15[3,4]									7250NI[3,4] 8410[3,4]
Z1Y	3111[3]		L254N[3,4] 254N[3,4] 254[3,4] L253[3,4] L254BK[3] L254SS[3,4] L067[3,4]		SC15[3,4]									7250NI[3,4] 8410[3,4]
Z2	3111[3,4]		L254N[3,4] 254N[3,4] 254[3,4] L253[3,4] L254BK[3,4] L254SS[3,4] L067[3,4]		SC15[3,4]									7250NI[3,4] 8410[3,4]
Z8			250[3,4]		Trey Combs Big Game[3,4]				3407SS[3,4] 34009[4]		811S[4]			
Z10	3111[3]		L254N[3,4] 254N[3,4] 254[3,4] L253[3,4] L254BK[3] L254SS[3,4] L067[3,4]		SC15[3,4]									7250NI[4] 8410[3,4]
Z11					C13U[3]			B420[3,4]						

Hooks that are ***bold italicized*** are barbless.
[1]Partridge considers this hook to be similar. The brand at the top of the column may or may not concur.
[2]The brand at the top of the column considers the Partridge hook to the left to be similar. Partridge may or may not concur.
[3]Similar in all aspects except the bend. See the corresponding characteristic tables.
[4]Similar in all aspects except the finish. See the corresponding characteristic tables.
[5]Similar in all aspects except the point position. See the corresponding characteristic tables.

Partridge	Daiichi	Dai-Riki	Eagle Claw	Gaelic Supreme	Gamakatsu	Hayabusa	Heritage	Kamasan	Mustad	Orvis	TMC	Turrall	Varivas	VMC
Z12	3111[3]		L254N[3,4] 254N[3,4] 254[3,4] L253[3,4] L254BK[3] L254SS[3,4] L067[3,4]		SC15[3,4]									7250NI[3,4] 8410[3,4]
Z13														
Z14														
Z15									77660SS[3,4]					
Z15XL														
Z17														
Z18														
Z30														
Z31														
Z33														
Z34														

Hooks that are ***bold italicized*** are barbless.
[1]Partridge considers this hook to be similar. The brand at the top of the column may or may not concur.
[2]The brand at the top of the column considers the Partridge hook to the left to be similar. Partridge may or may not concur.
[3]Similar in all aspects except the bend. See the corresponding characteristic tables.
[4]Similar in all aspects except the finish. See the corresponding characteristic tables.
[5]Similar in all aspects except the point position. See the corresponding characteristic tables.

PARTRIDGE Characteristics

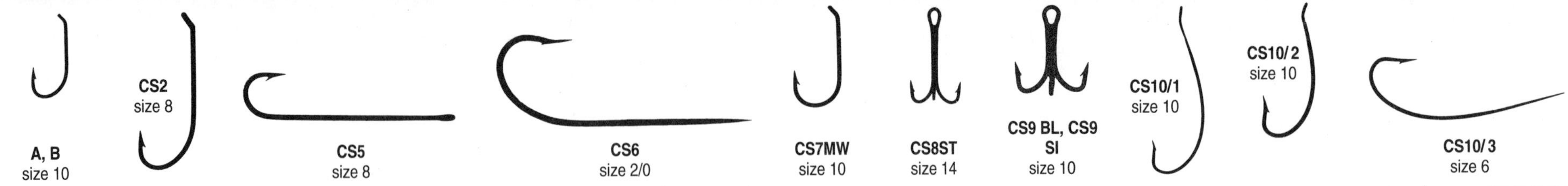

Hook	Use	Wire	Shank	Eye	Bend	Finish	Barb	Point	Common Name	Remarks	Size
A	Wet	S	1XS	TD	Spr-WG-R	Bron			Albert Partridge Wide Gape Down Eye	Nymphs. Same as B except has TD eye.	8-16
B	Hook Discontinued	S	1XS	TU	Spr-WG-R	Bron			Albert Partridge Wide Gape Up Eye	Same as A but with a TU eye.	10-16
CS2	Streamer	S	2XL	TD-L	Lim-M	Blk			S.E.B. Wet Fly	Forged.	2-8
CS5	Streamer	2XH	2XL	S	P-M	Blue			Keith Fulsher Thundercreek	Thunder Creek flies. Special bend.	4-8
CS6	Display	1XH	S	No	Lim-M	Blk			Adlington and Hutchinson Blind Eye Salmon	Traditional salmon bend. Detailed with hand-filed point.	8/0-2/0
CS7MW	Wet	S	S	TD	P-WG	Blk			Captain Hamilton International Middleweight	Forged.	10-14
CS8ST	Treble	1XF	2XL	S	N.I.-Treb	Blk			AB Sea Trout Treble	Saltwater, salmon.	12-16
CS9 BL	Treble	2XH	S	S	N.I.-Treb	Blk		Out	Rob Wilson XX Strong Outpoint Treble	Tube flies. Same as CS9 SI. Use with Waddington shank.	2-10
CS9 SI	Treble	2XH	S	S	N.I.-Treb	Sil		Out	Rob Wilson XX Strong Outpoint Treble	Tube flies. Same as CS9 BL. Use with Waddington shank.	2-10
CS10/1	Steelhead/ Salmon	S	2XL-C	TU-L	Dub	Blk		DP	Bartleet Traditional Fly	Slightly humped shank. Forged.	3/0-10
CS10/2	Steelhead/ Salmon	1XH	1XL	TU-L	Dub	Blk		DP	Bartleet Supreme Fly	Slightly humped shank. Forged.	1/0-10
CS10/3	Display	1XF	2XL-C	No	Dub	Blk			Bartleet Traditional Blind Eye	Forged.	3/0, 1/0, 2, 6

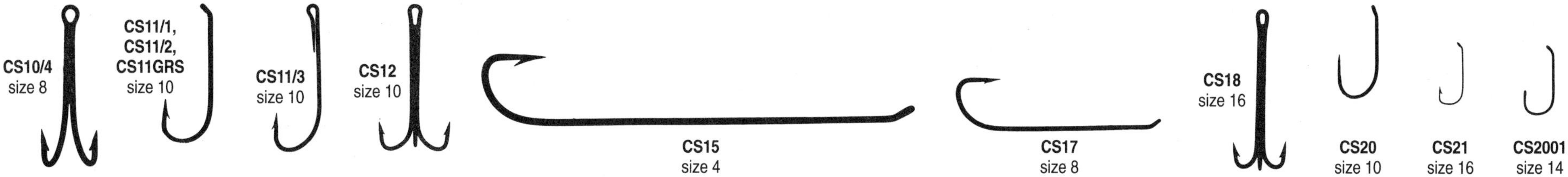

Hook	Use	Wire	Shank	Eye	Bend	Finish	Barb	Point	Common Name	Remarks	Size
CS10/4	Double	1XH	1XL	TU-L	Dub-Dbl	Blk			Bartleet Double	Dbl. version of CS10/2.	4-8
CS11/1	Streamer	S	4XL	TD	P-M	GS		Chem	JS Sea Streamer (Grey Shadow Finish)	Saltwater. Same as CS11/2 & CS11GRS. Redditch bend. Forged.	2-10
CS11/2	Streamer	S	4XL	TD	P-M	SS			JS Sea Streamer (Stainless Steel)	Saltwater. Same as CS11/1 & CS11GRS. Redditch bend. Forged.	4-8
CS11/3	Streamer	S	2XL	TD-L-M	Dub-M	SS	Out		JS Sea Streamer Supreme (Stainless Steel)	Eye loop taper is under the shank. Salmon bend. Forged.	1-10
CS11GRS	Streamer	S	4XL	TD	P-M	GS-(Rmks)		Chem	JS Sea Streamer (Grey Shadow)	Saltwater. Same as CS11/1 & CS11/2. Redditch bend available in SS w/ special order. Forged.	2-10
CS12	Treble	S	6XL	TU-L	N.I.-Treb	Blk			A.E.M. 6XL Shank Salmon Fly Treble		2-12
CS15	Streamer	1XH	10XL	TD-L	Lim	Bron			Carrie Stevens 10X Long Heavyweight Streamer	Pike flies. Forged.	2/0, 2, 4
CS17	Streamer	2XH	6XL	TD-L	Lim	Blk			Ken Baker Extra Long Limerick Streamer	Forged.	1-8
CS18	Treble	S	10XL	S	N.I.-Treb	Blk			Bill CEO Extra Long Sea Trout Treble	Saltwater, salmon. Used in place of Waddington shank.	12-16
CS20	Dry	1XF	1XS	TD	Spr-WG	Blk	No Barb	Spec	Mike Weaver Arrowpoint Barbless Dry Fly	Forged.	10-18
CS21	Dry	4XF	S	TD	Sneck	Blk			Darrel Martin Smut	Forged.	16-18
CS2001	Dry	4XF	S	TD	P	Bron	No Barb	Tag	Bramley/Betts Dry Fly Tag	Eyed point. Forged.	14-16

PARTRIDGE — Characteristics

Hook	Use	Wire	Shank	Eye	Bend	Finish	Barb	Point	Common Name	Remarks	Size
CS2002	Streamer	S	4XL	TD	P	Bron	No Barb	Tag	Bramley/Betts Longshank Tag	Eyed point. Forged.	8-10
CS2003	Nymph	2XF	2XL	S	N.I.-M	Blk	No Barb	Tag	Bramley/Betts Nymph Tag	Eyed point. Special bend. Forged.	12-14
CS27GRS	Dry	2XF	S	TD	N.I.-M	GS	No Barb	Chem	Roman Moser Barbless Dry Fly	Wets, emergers, dries. Special bend. Forged.	10-18
CS28GRS	Nymph	1XF	1XL	TD	N.I.-M	GS	No Barb	Chem	Roman Moser Barbless Nymph	Large dry flies. Special bend. Forged.	8-14
CS29GRS	Streamer	S	6XL	S	N.I.-M	GS	No Barb	Chem	Roman Moser Barbless Streamer	Special bend. Replaced CS29. Forged.	2-8
CS41	Bass	S	S	S	P-M	Bron		Chem	Bass "Flashpoint" Fly	Deer Hair. Redditch bend. Forged.	3/0, 1/0, 2, 6
CS42	Steelhead/ Salmon	1XF	1XL	TD-L	Lim-M	Bron			MW Bomber	Dry steelhead/salmon flies, streamers. Forged.	2-6
CS51	Saltwater	S	S	S	P	Tin-Sil			Homosassa Tarpon Saltwater Fly		5/0-2/0
CS52	Saltwater	2XH	S	S	Lim	SS			Sea Prince Saltwater Fly (Stainless Steel)	Tarpon & bonefish. Same as Z8.	6/0-10
CS53	Saltwater	2XH	3XL	S	Lim	SS			Sea Lord Saltwater Fly (Stainless Steel)	Bluefish & bonefish.	4/0-6
CS54	Saltwater	S	S	S	Lim-C	GS		Chem	Partridge Saltwater Shrimp Fly	Shrimp patterns and short streamers.	2-10
D3ST	Streamer	1XH	4XL	S	Spr-M	Bron			Partridge Straight Eye Streamer	Nymphs. Previously called the Lure Hook. Flattened bend. Forged.	4-14

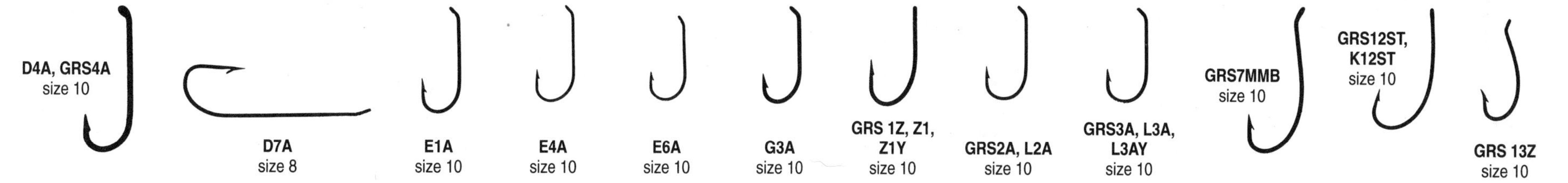

Hook	Use	Wire	Shank	Eye	Bend	Finish	Barb	Point	Common Name	Remarks	Size
D4A	Streamer	1XH	4XL	TD	P	Bron			Partridge Bucktail/Streamer	Bucktails, nymphs. Same as GRS4A. Forged.	2-14
D5B	Dry	2XF	2XL	TU	P-M-R	Bron	Mini		Partridge Mayfly (Up eye)	Forged.	8-14
D7A	Streamer	S	6XL	TD-M	Lim-M	Bron			Finnish Streamer	Special increased holding area (throat). Special bend. Forged.	2-8
E1A	Dry	4XF	1XL	TD	Spr-M	Bron			Hooper L/S 4X Fine Dry Fly	Nymphs. Redditch bend. Forged.	10-22
E3AY	Dry	4XF	S	TD	Spr-R	Bron	No Barb		Ritz Dry Fly (Barbless)	Forged.	10-18
E4A	Dry	4XF	S	TD	Spr-R	Bron			Ritz Dry Fly	Forged.	10-18
E6A	Dry	4XF	1XS	TD	Spr-M	Bron			Hooper 1X Short 4X Fine Dry Fly	Redditch bend. Forged.	8-18
G3A	Wet	S	S	TD	Spr	Bron			Sproat Forged Wet Fly	Nymphs. Forged.	6-16
GRS 1Z	Carp	S	S	S	Lim-M	GS		Chem	Grey Shadow Jack Hilton Carp	Same as Z1 & Z1Y. Jack Hilton style bend. Flattened bend. Forged.	2-10
GRS2A	Wet	2XF	S	TD	P-WG	GS		Chem	Grey Shadow Captain Hamilton Wet Fly	Dry flies. Same as L2A. Forged.	10-16
GRS3A	Dry	4XF	S	TD	P-WG	GS		Chem	Grey Shadow Captain Hamilton Dry Fly	Same as L3A & L3AY. Forged.	12-18
GRS4A	Streamer	1XH	4XL	TD	P	GS		Chem	Grey Shadow Bucktail/Streamer	Bucktails. Same as D4A.	2-10
GRS7MMB	Nymph	S	6XL-C	TU	C	GS		Chem	Jardine Living Nymph		8-20
GRS12ST	Nymph	S	3XL	S	C	GS		Chem	Grey Shadow Emerger/Nymph	Caddis, emergers. Same as K12ST.	8-14
GRS 13Z	Carp	1XH	S-C	TU-M	Dub-M	GS		Out-C	Kevin Maddocks Cassien Carp	Shrimp. Forged.	2-10

PARTRIDGE Characteristics

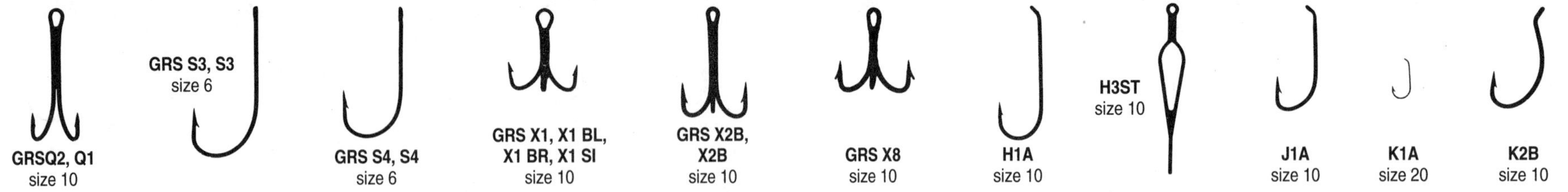

Hook	Use	Wire	Shank	Eye	Bend	Finish	Barb	Point	Common Name	Remarks	Size
GRSQ2	Double	1XH	2XL	TU-L	Lim-M-Dbl	GS			Grey Shadow Double Low Water	Same as Q1.	6-10
GRS S3	Saltwater	S	6XL	S	N.I.-M	GS		Chem	John Holden Middleweight Sea Fly (Grey Shadow)	Streamers, pike. Special bend. Same as S3. Forged.	6/0-6
GRS S4	Saltwater	2XH	5XL	S	N.I.-M	GS		Chem	John Holden Heavyweight Sea Fly (Grey Shadow Finish)	Streamers, pike. Special Bend. Same as S4. Forged.	6/0-6
GRS X1	Treble	1XH	S	S	N.I.-Treb	GS		Out-C	Partridge Outpoint Treble (Grey Shadow)	Same as X1 BR, X1 BL & X1 SI. Use with Waddington shank. Forged.	3/0-10
GRS X2B	Treble	2XH	1XL	TU-L	N.I.-Treb	GS		Out	Grey Shadow L/S Salmon Fly Treble	Salmon flies. Same as X2B.	6-10
GRS X8	Treble	S	1XL	S-U	N.I.-Treb	GS	Out	Chem	Hugh Faulkus Holdfast Outbarb Treble	Tube flies and hairwing salmon.	4-10
H1A	Nymph	1XF	2XL	TD	P-WG	Bron			Captain Hamilton Nymph	Streamers, dry flies, terrestrials, Muddlers. Forged.	2-14
H1B/CA	Nymph	S	3XL-BU	TU-M	P-M	Rmks			Oliver Edwards Correct Attitude Nymph	Streamers, larvae. 30° TU Eye. Shank is BU 8°. Available in bron. or light olive.	2-18
H3ST	Nymph/ Draper	S	6XL-S	S-L	Spr	Bron			Draper Flat-Bodied Nymph	Shank is split lengthwise for wide, flat bodies.	6-16
J1A	Wet	S	S	TD	Lim	Bron			Partridge Limerick Wet Fly		4-16
K1A	Midge	4XF	S	TD	P-WG-R	Bron			Vince Marinaro Midge	Forged.	20-28
K2B	Nymph	1XH	S-C	TU	C-WG	Bron			Yorkshire Sedge/Caddis	Forged.	8-18

Hook	Use	Wire	Shank	Eye	Bend	Finish	Barb	Point	Common Name	Remarks	Size
K3A	Dry	4XF	2XL-BU	TD	Spr-M	Bron			Swedish Dry Fly	Shank has special kink after eye for upside-down flies. Forged.	10-18
K4A	Scud/Shrimp	S	1XS-C	TD	C-R	Bron			John Veniard Grub/Shrimp	Emergers. Forged.	2-18
K6ST	Nymph	1XF	3XL-BU	S	P	Bron			Taff Price Swimming Nymph	Swimming nymphs. Forged.	8-14
K7	Dry	S	1XS-BU	TU	P	Bron			Short Shank Mayfly	Designed by Jens Pilgaard.	10, 12
K10	Dry	4XF	Rmks	F-L	P-M	Bron			Yorkshire Fly Body	Shank is looped over top and extends past bend, then curves upward. Redditch bend. Forged.	10-16
K12ST	Nymph	S	3XL	S	C	Bron			Long Shank Sedge/Caddis	Caddis, emergers. Same as GRS12ST.	8-22
K14ST	Nymph	S	2XL-C	S	York	Sil			Oliver Edwards Nymph/Emerger	Emergers, dry flies. Forged.	4-20
K16B	Dry	S	S-Rmks	TU	Spr-M	Bron			Long Shank Mayfly	Mayflies, nymphs. Shank has distinct kink at thorax. Designed by Jens Pilgaard.	10, 12
K17	Dry	S	S	F-M	P	Bron			Davy Wotton In Line+Dry Fly Nymph Emerger Hook	Nymphs, emergers, midges. Mfg. states "In line eye."	12-26
L2A	Wet	2XF	S	TD	P-WG	Bron			Captain Hamilton Wet Fly	Dry flies. Same as GRS2A. Forged.	6-18
L3A	Dry	4XF	S	TD	P-WG	Bron			Captain Hamilton Dry Fly (Down Eye)	Same as L3AY & GRS3A. Forged.	6-22
L3AY	Dry	4XF	S	TD	P-WG	Bron	No Barb		Captain Hamilton Barbless Dry Fly	Same as L3A & GRS3A. Forged.	10-16
L3B	Dry	4XF	S	TU	P-WG	Bron			Captain Hamilton Dry Fly (Up Eye)	Forged.	10-18

Characteristics

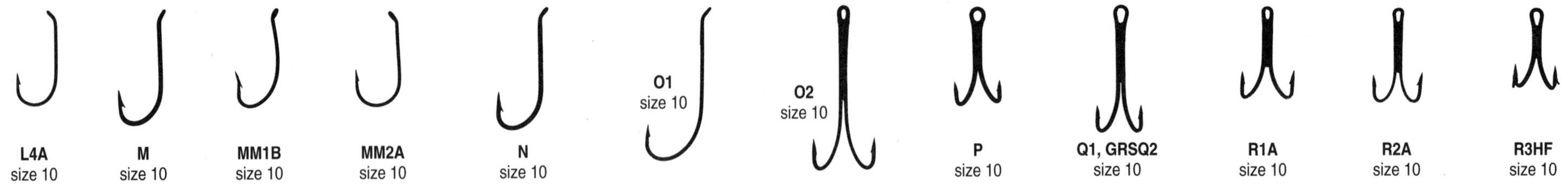

Hook	Use	Wire	Shank	Eye	Bend	Finish	Barb	Point	Common Name	Remarks	Size
L4A	Dry	6XF	S	TD	P	Bron			Captain Hamilton Featherweight Dry Fly	Forged.	10-20
M	Steelhead/ Salmon	2XH	S	TU-L	Dub-M	Blk			Partridge Single Salmon	Saltwater. Salmon bend. Forged.	5/0-10
MM1B	Wet	S	1XS	TU	Lim	Blk		Spec	McHaffie Masters Wet Fly	Special Dublin point.	2-12
MM2A	Wet	S	1XS	TD	P-WG	Bron			Bob Church Goldhead	Nymphs, beadheads. Designed to easily attach beads. Forged.	8-14
MM3ST	Tube	S	S	S	Spr-M	Blk			Nordic Single Tube Fly		2-10
N	Steelhead/ Salmon	1XH	3XL	TU-L	Dub-M	Blk			Partridge Single Low Water	Saltwater. Salmon bend. Forged.	8/0-10
O1	Steelhead/ Salmon	S	6XL	TU-L	Wil	Blk			Partridge Single Wilson	Dry Steelhead/salmon flies, saltwater. Forged.	2-16
O2	Double	S	6XL	TU-L	Wil-Dbl	Blk			Partridge Double Wilson	Saltwater.	2-16
P	Double	2XH	S	TU-L	Lim-M-Dbl	Blk			Partridge Double Salmon	Steelhead/salmon wet flies. Salmon bend.	2-12
Q1	Double	1XH	2XL	TU-L	Lim-M-Dbl	Blk			Partridge Double Low Water	Same as GRSQ2.	3/0-12
R1A	Double	S	1XL	TD-L	Lim-Dbl	Bron			Partridge Double Limerick	Steelhead/salmon, saltwater, wet flies.	2-12
R2A	Wet	S	S	TD-L	P-M-Dbl	Bron	Out		Partridge Outpoint Double	Saltwater, steelhead/ salmon. Redditch bend. Forged.	10-18
R3HF	Double	2XH	2XS	S-L	N.I.-Dbl	Blk	Out		Hugh Faulkus Holdfast Outbarb Double	Steelhead/salmon, saltwater, wet flies. Special bend. Forged.	6-14

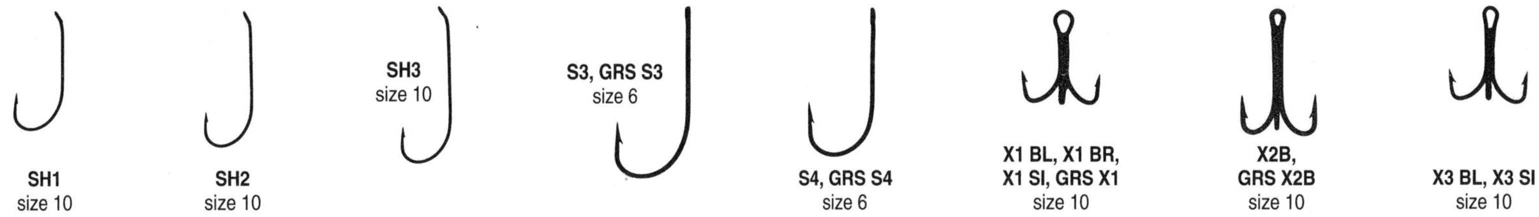

Hook	Use	Wire	Shank	Eye	Bend	Finish	Barb	Point	Common Name	Remarks	Size
SH1	Wet	2XH	S	TD	Lim	GS		Chem	Stronghold Wet Fly	Nymphs. Forged.	8-16
SH2	Nymph	1XH	1XL	TD	Lim	GS		Chem	Stronghold Nymph	Wet flies. Size 12 is within international competition regulation limits. Forged.	8-16
SH3	Streamer	S	3XL	TD	Spr	GS		Chem	Stronghold Long Shank	Nymphs. Size 14 is within international competition regulation limits.	8-16
S3	Saltwater	S	6XL	S	N.I.-M	GS		Chem	John Holden Heavyweight Sea Hooks	Streamers, pike. Special bend. Same as GRS S3. Forged.	8/0-6
S4	Saltwater	2XH	5XL	S	N.I.-M	GS		Chem	John Holden Heavyweight Sea Hooks	Streamers, pike. Special bend. Same as GRS 4S. Forged.	6/0-6
X1 BL	Treble	1XH	S	S	N.I.-Treb	Blk		Out	Partridge Outpoint Treble (Black)	Same as X1 BR, X1 SI & GRS X1. Use with Waddington shank. Forged.	3/0-16
X1 BR	Treble	1XH	S	S	N.I.-Treb	Bron		Out	Partridge Outpoint Treble (Bronze)	Same as X1 BL, X1 SI & GRS X1. Use with Waddington shank. Forged.	3/0-16
X1 SI	Treble	1XH	S	S	N.I.-Treb	Sil		Out	Partridge Outpoint Treble (Silver)	Same as X1 BR, X1 BL & GRS X1. Use with Waddington shank. Forged.	3/0-16
X2B	Treble	2XH	1XL	TU-L	N.I.-Treb	Blk		Out	Partridge L/S Salmon Fly Treble	Salmon flies. Same as GRS X2B.	2-16
X3 BL	Treble	S	S	S-U	N.I.-Treb	Blk			Partridge Needle Eye Tube Fly Treble (Black)	Tube flies. Same as X3 SI.	2-16
X3 SI	Treble	S	S	S-U	N.I.-Treb	Sil			Partridge Needle Eye Tube Fly Treble (Silver)	Tube flies. Same as X3 BL.	4-12

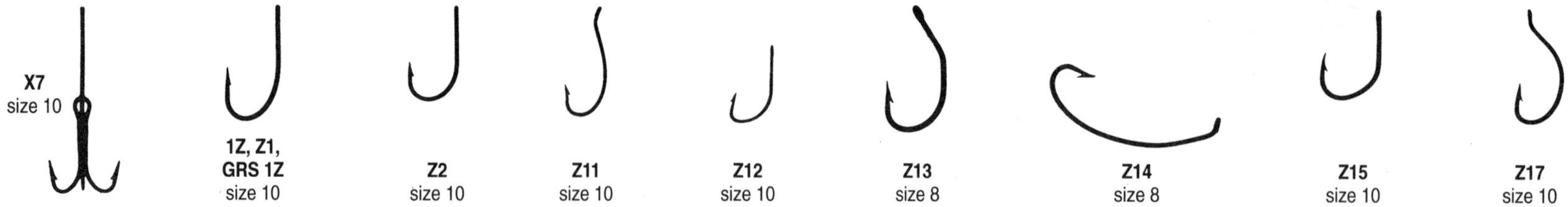

Hook	Use	Wire	Shank	Eye	Bend	Finish	Barb	Point	Common Name	Remarks	Size
X7	Treble	S	S-Rmks	S	N.I.-Treb	Blk			Partridge Needle Eye Tube Fly Treble	Tube flies. Shank has extended spike to fit into tubes of tube flies.	4-12
Z1	Carp	S	S	S	Lim-M	Blk		Chem	Jack Hilton Carp Hook	Same as Z1Y & GRS 1Z. Jack Hilton style bend. Flattened bend. Forged.	1/0-12
Z1Y	Carp	S	S	S	Lim-M	Blk	No Barb	Chem	Jack Hilton Barbless Carp Hook	Same as Z1 & GRS 1Z. Jack Hilton style bend. Flattened bend. Forged.	1-10
Z2	Carp	S	S	S	WG-S-R	Bron		Chem	Partridge Record Breaker Specialist Hook	Forged.	1/0-18
Z8	Saltwater	2XH	S	S	Lim	SS			"Sea Prince" Saltwater Fly	Tarpon, bonefish. Same as CS52.	6/0, 4/0-4
Z10	Saltwater	S	S	S	Aber	Blk	Mini		Partridge Aberdeen	Bass, pike.	6/0-8
Z11	Carp	1XF	S-C	TU-M	Dub-M	Blk		Out-C	Kevin Maddocks Hair Rig Hook	Scuds & shrimp. Forged.	2-12
Z12	Carp	S	S	S	Dub-M-R	Blk	Out	Chem	Kevin Maddocks Outbarb Specimen Hook	Forged.	2-10
Z13	Carp	1XH	S-BD	S	Spr-M	GS		Chem	Ritchie McDonald Carp Hook	Eye and point are in line with each other. Forged.	2-8
Z14	Carp	S	S-C	TD	S	Blk		Out-C	Partridge "Piggyback" Bent Hook	Shank is uniquely curved. Eye and point are in line with each other. Forged.	2-8
Z15	Carp	1XH	1XS	S	WG-M	Blk			Kevin Maddocks Boilie Carp Hook	Forged.	2-10
Z15XL	Carp	1XH	1XL	S	WG-M	GS			Kevin Maddocks Boilie Carp Hook	Forged.	2-10
Z17	Carp	1XH	S-C	S	WG-R	Bron			WS Specimen Hunter	Eye and point are in line with each other. Forged.	1-10

Z18
size 8

Z30
size 10

Z31
size 10

Z33
size 8

Z34
size 8

Hook	Use	Wire	Shank	Eye	Bend	Finish	Barb	Point	Common Name	Remarks	Size
Z18	Carp	1XxH	S-C	S	WG-R	Bron	Rmks	Chem	WS Barbless Arrowpoint Specimen Hunter Hook	Eye and point are in line with each other. Point has flat top and bottom and looks like a spade. Forged.	2-8
Z30	Carp	1XH	1XL	S	WG-S-R	Bron		Chem	Dutch Specialist Carp	Forged.	2/0-10
Z31	Carp	1XH	1XL	S	WG-S-R	Blue	No Barb	Chem	Dutch Barbless Blue	Forged.	4-10
Z33	Carp	2XH	S	TU	S-R	GS		Chem	Rod Hutchinson WW Carp Hook	Jack Hilton style bend. WW means weedy waters. Forged.	2-8
Z34	Carp	1XH	S	TU	P-M-R	Bron		Chem	Rod Hutchinson OW Carp Hook	OW means open waters. Forged.	2-8

Grey Shadow is a trademark of Partdige of Redditch, Ltd.

TIEMCO (TMC) Cross-Reference

TMC	Daiichi	Dai-Riki	Eagle Claw	Gaelic Supreme	Gamakatsu	Hayabusa	Heritage	Kamasan	Mustad	Orvis	Owner	Partridge	Varivas	VMC
100	1100[2,3]	300[2,3] 305[2,3]	***L061B***[2,3] L059[3] L060[3]		S10-3F[3,4]	572[3]		B401[2] B400[3]	94840[3] AC94840[3] ***94845***[3] 80000BR[3] ***80010BR***[3]	1509[2,3] 1523[3]		L3A[3] ***L3AY***[3,4] GRS3A[3,4]		V7061[3] 9280[3]
100BL	1100[2,3]	300[2,3] 305[2,3]	***L061B***[2,3] L059[3] L060[3]		S10-3F[3,4]	572[3]		B400[3]	94840[3] AC94840[3] ***94845***[3] 80000BR[3] ***80010BR***[3]	1509[2,3] 1523[3]		L3A[3] ***L3AY***[3,4] GRS3A[3,4]		V7061[3] 9280[3]
101	1110[2,3]	310[2,3]			S10S[2] S-13S-M[3] S12S-1F[3]			B410[2,3]	94859[3]	170T[3,4]				
102Y	1100[3,4]		***L061B***[3,4] L059[3,4] L060[3,4]		S10-3F[3]	572[3]		B400[3,4]	94840[3,4] AC94840[3,4] ***94845***[3,4] 80000BR[3,4] ***80010BR***[3,4]	1523[3,4]		L3A[3,4] ***L3AY***[3,4] GRS3A[3,4]		V7061[3,4] 9280[3,4]
103BL	1100[3,4]		***L061B***[3,4] L059[3,4] L060[3,4]		S10-3F[3]	572[3]			94840[3,4] AC94840[3,4] ***94845***[3,4] 80000BR[3,4] ***80010BR***[3,4]	***1877***[2,3,4] 1523[3,4]		L3A[3,4] ***L3AY***[3,4] GRS3A[3,4]		V7061[3,4] 9280[3,4]
105	2571[2]				C14S[2]									
109BL														
146Z	1100[4]		***L061B***[4] L059[4] L060[3,4]		S10-3F[4]	572[3]		B400[3,4]	94840[4] AC94840[4] ***94845***[4] 80000BR[3,4] ***80010BR***[3,4]	1523[4]		L3A[3,4] ***L3AY***[3,4] GRS3A[3,4]		V7061[4] 9280[3,4]
200R	1270[2,3,5] 1273[3,4,5]	270[2,3,5]	L052[2,3,5]			383[3,5]		B220[2,3,5]		1510[2,3,5]				
200RBL	1270[2,3,5] 1273[3,4,5]	270[2,3,5]	L052[2,3,5]			383[3,5]		B220[3,5]		1510[2,3,5]				

Hooks that are ***bold italicized*** are barbless.
[1] TMC considers this hook to be similar. The brand at the top of the column may or may not concur.
[2] The brand at the top of the column considers the TMC hook to the left to be similar. TMC may or may not concur.
[3] Similar in all aspects except the bend. See the corresponding characteristic tables.
[4] Similar in all aspects except the finish. See the corresponding characteristic tables.
[5] Similar in all aspects except the point position. See the corresponding characteristic tables.

TMC	Daiichi	Dai-Riki	Eagle Claw	Gaelic Supreme	Gamakatsu	Hayabusa	Heritage	Kamasan	Mustad	Orvis	Owner	Partridge	Varivas	VMC
205BL			L144[2,3,5]	Caddis[3,5]	C12U[2,3,5]	*388*[3]			37160[3,5] 37161[3,4,5] 37162[3,4,5]					
206BL	1140[2]													
246Z														
300	2340[2,3] 2220[2]				Russian River[3]					1511[2,3]				
346Z														9283[3]
400T	1770[2,3]					380			80150BR[3]	1512[2,3]		K6ST[3]		
411S														
500U					S10U[2,4]			B440[2]						
501					S13S-M[2]								988[3]	
511S			L200[3,4]						90233S[3] 34015D[3,4] 33900[3,4]					
531														
700			L1197B[3,4,5] L1197G[3,4,5] L1197N[3,4,5] NT1197[3,4,5]					B175[3,4,5]		167T[3,4,5]				
800B Discontinued		930[4,5]	L054SS[4,5] L7254[4,5]		L11S-3H[2,3,4] S14S-3H[2,3,4]						5170[3,4,5]			
800S		930[5]	L067[2,3,4,5] L054SS[5] L7254[4,5]		L11S-3H[2] Trey Combs Big Game						5170[3,4,5]			
811S	2546[2]	930[2]	L054SS[2] L7254[4]		SS15[2,3,4]						5170[3,4]			

TIEMCO (TMC) Cross-Reference

TMC	Daiichi	Dai-Riki	Eagle Claw	Gaelic Supreme	Gamakatsu	Hayabusa	Heritage	Kamasan	Mustad	Orvis	Owner	Partridge	Varivas	VMC
900BL	1100[3,4,5]		L059[3,4,5] L060[3,4,5] ***L061B***[3,4,5]		S10-3F[3,5]	572[3,5]		B402[2,3,4] B400[3,4,5]	94840[3,4,5] AC94840[3,4,5] ***94845***[3,4,5] 80000BR[3,4,5] ***80010BR***[3,4,5]	1523[3,4,5]		L3A[3,4,5] ***L3AY***[3,4,5] GRS3A[3,4,5]		9280[3,4,5] V7061[3,4,5]
902BL									3399D[3,4]			L2A[3,4] ***CS27GRS***[3,4] GRS2A[3,4]	98i[3]	V7060[3,4]
905BL			***L064B***[3,4]											
911S														
921	1310[2]				S10-2S[2]	751[3]		B405[3]	94838[3]				981[3]	
947BL	1220[3] 1222[3,4]													
2302														
2312	1270[3] 1273[3,4]		L052											
2457	1150[2] 1155[2]	135[2]	L056[2]	Grub[3,4]		387[3]	Grub[3]	B110[3]	80200BR[3]					
2487	1130[2]		L055[2,3]		C12[2]			B100[2,3]		1639[2,3]		K4A[3]		
2487BL	1130[2]		L055[2,3]		C12[2]			B100[3]		1639[2,3]		K4A[3]		
2487G	1130[3,4]		L055[3,4]		C12[4]			B100[3,4]		1639[3,4]		K4A[3,4]		
2488														
3761	1560[2]	060[2]	L057[2]							167T[2]				V7071
3761BL	1560[2]	060[2]	L057[2]							167T[2]				V7071
3769	1530[2]	075[2] 070[2]			L10-2H[2]			B175[2]		1641[2]		SH1[3,4]	977[3,4]	

Hooks that are ***bold italicized*** are barbless.
[1] TMC considers this hook to be similar. The brand at the top of the column may or may not concur.
[2] The brand at the top of the column considers the TMC hook to the left to be similar. TMC may or may not concur.
[3] Similar in all aspects except the bend. See the corresponding characteristic tables.
[4] Similar in all aspects except the finish. See the corresponding characteristic tables.
[5] Similar in all aspects except the point position. See the corresponding characteristic tables.

TMC	Daiichi	Dai-Riki	Eagle Claw	Gaelic Supreme	Gamakatsu	Hayabusa	Heritage	Kamasan	Mustad	Orvis	Owner	Partridge	Varivas	VMC
5210	1180[2] 1170[2] 1182[4] 1100	305[2]	L059 L060[3] ***L061B***		P-10[2] S10[2,4] ***S10-B***[2,4] S10-3F[4]	572[3]		B400[3]	94840 AC94840 ***94845*** 80000BR ***80010BR***	1523[2]		E1A[3] L3A[3] ***L3AY***[3] GRS3A[3,4]		V7061 9280[3]
5212	1280[2]	730	L060[2]		P10-2L1H[2]	552				1638[2]		H1A[3]		V7062
5230					S10-3F[2,3]				94833			L4A		9288
5262	1710[2]	730[2]	L063[2]			376 ***381***			9671 AC9671	1524[2]				9279[3]
5263	1720[2]	710[2]	L281[2]		S11-4L2H[2,3]				9672 AC9672	1526[2]		H1A[3]		V7070[3]
5263BL	1720[2]	710[2]	L281[2]		S11-4L2H[2,3]				9672 AC9672	1526[2]		H1A[3]		V7070[3]
7931C				F-F Heavy Spey[3]										
7989	2421[2,3]			Single Dry Fly[3]	T10-3H[2,3]				9049[3]	1644[2,3]				
7999	2441[2]			Single Low Water[3]	T10-6H[2,3]			B180[2,3] B190[3]	36890[3]	1645[2,3]		M[3]	974	
8089	2720[2,3] 2722[2,3,4]									8810[2,3]				
8089NP	2720[3,4] 2722[3]									8810[3,4]				
9300		075	L059[2,3] L1197B[3] L1197G[3,4] L1197N[3,4] NT1197[3,4]		P10[2,3]			B175[3]		167T[3]				

TIEMCO (TMC) Cross-Reference

TMC	Daiichi	Dai-Riki	Eagle Claw	Gaelic Supreme	Gamakatsu	Hayabusa	Heritage	Kamasan	Mustad	Orvis	Owner	Partridge	Varivas	VMC
9394	1750[3,4]				S11S-4L2H[2,3]				9674[3,4] 9674A[3,4] 9674B[3,4]					
9395	1750[2,3]								9674[3] 9674A[3,4] 9674B[3,4]	0167[2]				

Hooks that are ***bold italicized*** are barbless.
[1] TMC considers this hook to be similar. The brand at the top of the column may or may not concur.
[2] The brand at the top of the column considers the TMC hook to the left to be similar. TMC may or may not concur.
[3] Similar in all aspects except the bend. See the corresponding characteristic tables.
[4] Similar in all aspects except the finish. See the corresponding characteristic tables.
[5] Similar in all aspects except the point position. See the corresponding characteristic tables.

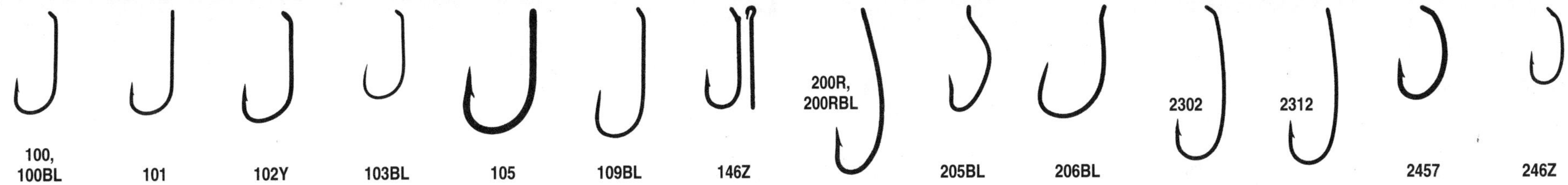

Hook	Use	Wire	Shank	Eye	Bend	Finish	Barb	Point	Remarks	Size
100	Dry	1XF	S	TD	P-M	Bron		Chem	Same as 100BL. Standard dry fly, slightly wider gape and slightly stronger wire strength in the smaller sizes. Forged.	8-26
100BL	Dry	1XF	S	TD	P-M	Bron	No Barb	Chem	Same as 100. Standard dry fly, slightly wider gape and slightly stronger wire strength in the smaller sizes. Forged.	10-22
101	Dry	1XF	S	S	P-M	Bron		Chem	Forged.	8-26
102Y	Dry	1XF	S	TD	Lim-M	Blk		Chem	Parachute flies. Odd-numbered sizes. Forged.	9-19
103BL	Dry	1XF	S	TD	P-M	Blk	No Barb	Chem	Odd-numbered sizes. Forged.	11-19
105	Egg	2XH	7XS	S	P-M-R	Bron		Chem	Tube flies. Forged.	4-10
109BL	Dry	1XF	3XL	TD	P-M	Blk	No Barb	Chem	Dry flies, wet flies, hoppers. Shank lengths 3XL, 1.5XL and 1XL dependent on size. Odd-numbered sizes. Forged.	7-19
146Z	Dry	1XF	S	TD-M	P	Blk		Chem	Goetz Quick Eye. Eye has an opening for easier tippet attachment. Forged.	10-22
200R	Nymph	S	3XL-C	S	Spr-M	Bron		SDP-C	Same as 200RBL. Dry flies. Gradual shank curve. Forged.	4-22
200RBL	Nymph	S	3XL-C	S	Spr-M	Bron	No Barb	SDP-C	Same as 200R. Dry flies. Gradual shank curve. Forged.	4-22
205BL	Scud/Shrimp	S	S-C	TU	Spr-M	Bron	No Barb	SDP-C	Caddis pupae, nymphs. Pronounced up shank curve. Mfg. states gape is 5x wider than standard.	8-20
206BL	Nymph	3XF	2XS-C	TU-M	Spr-M	Blk	No Barb	Chem	Nymphs, dry flies, scuds, CDC & emergers. Wire is variable to size 3XF to 1XF. Shank curve is gradual to make hook ride as a straight eye.	6-20
2302	Terrestrial	S	2XL-C	TD	Spr-M	Bron		Chem	Stoneflies, caddis & hoppers. Slight hump in shank. Same as 2312. Forged.	6-16
2312	Terrestrial	1XF	2XL-C	S	Spr-M	Bron		Chem	Stoneflies, caddis & hoppers. Slight hump in shank. Same as 2302. Forged.	
2457	Scud/Shrimp	2XH	2XS-C	TD	Spr-M	Bron		Chem	Caddis pupae, eggs. Mfg. states gape is 2x wider than standard. Forged.	6-18
246Z	Scud/Shrimp	2XH	S-C	TD-M	Spr-M	Blk		Chem	Goetz Quick Eye. Eye has an opening for easier tippet attachment. Forged.	10-20

TIEMCO (TMC) — Characteristics

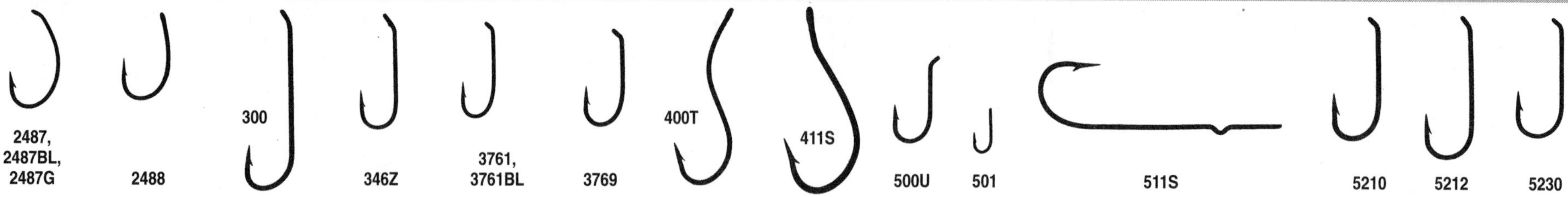

Hook	Use	Wire	Shank	Eye	Bend	Finish	Barb	Point	Remarks	Size
2487	Scud/Shrimp	1XF	2XS-C	TD	Spr-M	Bron		Chem	Caddis pupae, emergers. Mfg. states gape is 2x wider than standard. Same as 2487BL & 2487G. Forged.	10-24
2487BL	Scud/Shrimp	1XF	2XS-C	TD	Spr-M	Bron	No Barb	Chem	Caddis pupae, emergers. Mfg. states gape is 2x wider than standard. Same as 2487 & 2487G. Forged.	12-22
2487G	Scud/Shrimp	1XF	2XS-C	TD	Spr-M	Gold		Chem	Caddis pupae, emergers. Mfg. states gape is 2x wider than standard. Same as 2487 & 2487BL. Forged.	10-20
2488	Scud/Shrimp	1XF	2XS	S	C-WG	Bron		Chem	Caddis, pupae, emergers. Mfg. states gape is 3x wider than standard. Forged.	10-24
300	Streamer	1XH	6XL	TD	Spr	Bron		Chem	Classic streamer patterns. Forged.	2-14
346Z	Nymph	1XH	1XS	TD-M	P	Blk		Chem	Goetz Quick Eye. Eye has an opening for easier tippet attachment. Forged.	10-18
3761	Nymph	2XH	1XL	TD	Spr	Bron		Chem	Wet flies. Same as 3761BL. Forged.	2-20
3761BL	Nymph	2XH	1XL	TD	Spr	Bron	No Barb	Chem	Wet flies. Same as 3761. Forged.	10-18
3769	Nymph	2XH	S	TD	Spr	Bron		Chem	Wet flies. Forged.	6-18
400T	Nymph	1XF	S-BU	S	Spr-M	Bron		Chem	Swimming nymph. Forged.	8-14
411S	Saltwater	1XH	S-BU	S	Spr-M	SS		Chem	Special bend back design for snag-free fishing. Forged.	3/0, 1/0, 4-8
500U	Dry	S	2XS	TU	P	Bron		Chem	Tricos & midges. Straight Point. Forged.	16-22
501	Dry	S	1XS	S	P	Bron		Chem	Midges. Small sizes with standard wire. Forged.	10-24
511S	Popper	S	4XL-PK	S	P-M	SS		Chem	Saltwater. Mfg. states gape is 2x wider than standard. Forged.	2/0-1/0, 2, 6
5210	Dry	1XF	S	TD	P	Bron		Chem	Traditional dry flies. Forged.	10-20
5212	Terrestrial	1XF	2XL	TD	P	Bron		Chem	Hoppers, stoneflies, caddis, and long mayfly duns. Forged.	6-16
5230	Dry	3XF	S	TD	P	Bron		Chem	Forged.	10-18

Characteristics TIEMCO (TMC)

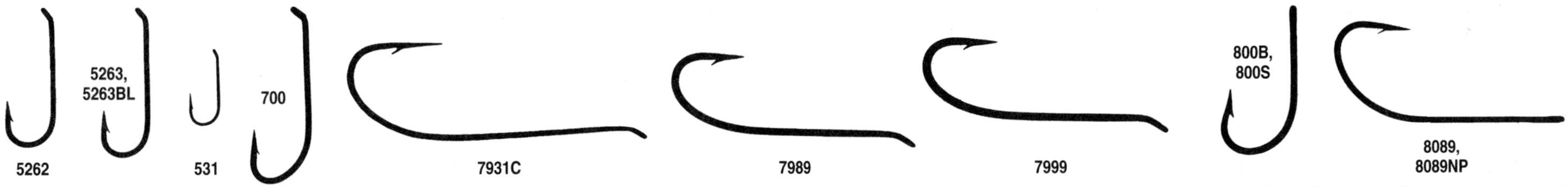

Hook	Use	Wire	Shank	Eye	Bend	Finish	Barb	Point	Remarks	Size
5262	Nymph	2XH	2XL	TD	P	Bron		Chem	Streamers. Forged.	2-18
5263	Nymph	2xH	3XL	TD	P	Bron		Chem	Same as 5263BL. Streamers. Forged.	2-18
5263BL	Nymph	2XH	3XL	TD	P	Bron	No Barb	Chem	Same as 5263. Streamers. Forged.	4-12
531	Dry	2XF	2XS	TD-M	P-M	Blk	Mini	Chem	Parachutes and thorax flies. Eye partially TD. Gape is slightly wide. Forged.	10-20
700	Steelhead/ Salmon	1XH	S	TD	Lim	Blk		SDP-C	Streamers, wet flies. Forged.	1/0-10
7931C	Steelhead/ Salmon	2XH	2XL	TU-L	Lim-WG	Blk				5/0-2/0
7989	Steelhead/ Salmon	1XF	S	TU-L	Spr-M	Blk		Chem	Steelhead/salmon dry flies. Forged.	2-8
7999	Steelhead/ Salmon	1XH	S	TU-L	Spr-M	Blk		Chem	Steelhead/salmon wet flies. Forged.	2/0-12
800B	Steelhead/ Salmon	1XH	S	S	Osh	Blk		SDP-C	Same as 800S. Forged. Hook no longer available.	N/A
800S	Saltwater	1XH	S	S	Osh	SS		SDP	Same as 800B. Tube flies. Forged.	4/0-8
8089	Bass	1XF	S	S	Lim-M	Bron		Chem	Bass bugs, deer-hair bugs. Same as 8089NP. Forged.	2, 6, 10
8089NP	Bass	1XF	S	S	Lim-M	Nick		Chem	Bass bugs, deer-hair bugs. Same as 8089. Forged.	2, 6, 10

TIEMCO (TMC) Characteristics

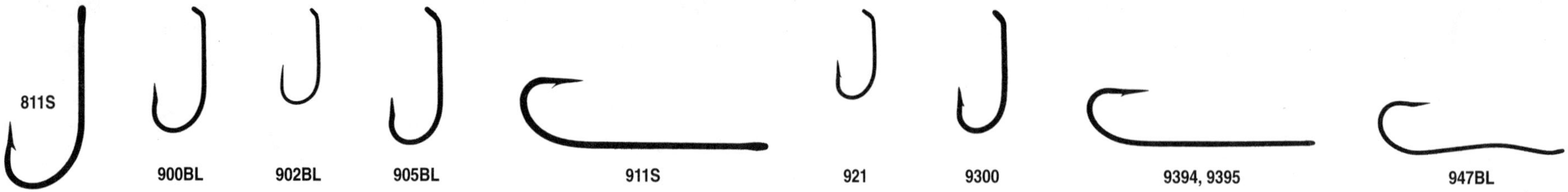

Hook	Use	Wire	Shank	Eye	Bend	Finish	Barb	Point	Remarks	Size
811S	Saltwater	1XH	S	S	Osh	SS		Chem	Tube flies. Forged.	4/0-8
900BL	Dry	1XF	S	TD	P-M	Blk	No Barb	SDP-C	Point goes outward and has a deeper throat than normal. Mfg. states gape is 1x wider than standard. Forged.	8-24
902BL	Dry	2XF	S	TD	P-M	Blk	No Barb	Chem	Sparse dries and emergers. Point slightly turned inward to compensate for fine wire.	12-20
905BL	Steelhead/ Salmon	1XF	S	TD	P-M	Blk	No Barb	Chem	Dry flies. Smaller sizes use standard wire. Forged.	2-10
911S	Saltwater	3XH	4XL	S	Spr	SS		Chem	Streamers. Forged.	5/0-4
921	Dry	1XF	2XS	TD	Spr	Bron		Chem	Parachutes, thorax flies, mayfly emerger, CDC & extended mayfly patterns. Forged.	8-20
9300	Wet	1XH	S	TD	Spr-M	Bron		Chem	Hairwing dry flies & soft hackle flies. Forged.	8-20
9394	Streamer	3XH	4XL	S	Osh	Nick		Chem	Saltwater, bass. Same as 9395.	2-10
9395	Streamer	3XH	4XL	S	Osh	Bron		Chem	Saltwater, bass. Same as 9394.	2-10
947BL	Nymph	S	3XL-BU	TD	P-M	Bron	No Barb	Chem	Nymphs, stoneflies, mayflies, dry flies & terrestrials. Forged.	4-20

Turrall	Daiichi	Dai-Riki	Eagle Claw	Gaelic Supreme	Gamakatsu	Hayabusa	Heritage	Kamasan	Mustad	Orvis	Partridge	Varivas	VMC
Barbless Extra-fine							Trout Xtra Fine				***CS2001***[5] CS21[3,4] ***E3AY***[3] E4A[3] GRS3A[3,4] L3A[3] ***L3AY***[3,4]		
Barbless Extra-long		710							9672 AC9672 38941[3]	1526	SH3[3,4]		
Barbless-Sproat	1550	070[3]	L057		S10[3,4]	373			3906 3399 3399A 3399N[4] 7948A[3]		G3A J1A[3] CS7MW[4]	98r[3,4]	8526
Bead Streamer		710[3]							9672[3] AC9672[3] 38941[3]	1526[3]	SH3[3,4]		
Bead Extra-long Nymph		300[3]						B170[3]	3906B[3] AC3906B[3] 7957B[3]				8527[3]
Bead Standard Nymph	1550[3]	070[3]	L057[3]		S10[3,4]	373[3]			3906[3] 3399[3] 3399A[3] 3399N[4] 7948A[3]		G3A[3] J1A[3] CS7MW[4]	98r[3,4]	8526[3]
Extra-fine Dry Fly							Trout Xtra Fine				***CS2001***[3] CS21[3,4] ***E3AY***[3] E4A[3] GRS3A[3,4] L3A[3] ***L3AY***[3,4]		
Extra-long Nymph		300						B170[3]	3906B[3] AC3906B[3] 7957B				8527[3]

Hooks that are ***bold italicized*** are barbless.
[1] Turrall considers this hook to be similar. The brand at the top of the column may or may not concur.
[2] The brand at the top of the column considers the Turrall hook to the left to be similar. Turrall may or may not concur.
[3] Similar in all aspects except the bend. See the corresponding characteristic tables.
[4] Similar in all aspects except the finish. See the corresponding characteristic tables.
[5] Similar in all aspects except the point position. See the corresponding characteristic tables.

Turrall	Daiichi	Dai-Riki	Eagle Claw	Gaelic Supreme	Gamakatsu	Hayabusa	Heritage	Kamasan	Mustad	Orvis	Partridge	Varivas	VMC
Extra-long Streamer		710					Trout Longshank		9672 AC9672 38941[3]	1526	SH3[3,4]		
Grub			L056[3]		C12								9902[3]
Keel Dry Fly													
Mayfly											E1A[3]		
Midge									94833				
Salmon Blind Eye				F-F Blind Eye									
Salmon Double	7131			Double Low Water[3] Double Wet				B280[3]			Q1[3] GRSQ2[3,4]		9410[4]
Salmon-Double Low-water				Double Low Water[3] Double Wet				B280[3]	80525BL		Q1[3] GRSQ2[3,4]		9410[4]
Salmon Extra-fine Dry Fly				F-F Light Wire[3] Single Dry Fly[3]					90240				
Salmon Low-water				F-F Low Water[3]			Single Salmon Low Water[3]		80500BL				
Salmon Single	2421[3]					771[3]		B180[3]	36890				
Salmon Treble				Treble Low Water[3,5] Treble Wet[3,5]				B380[3,5]			X2B[3] GRSX2B[3,4] CS12[3]		
Salmon Treble Low-water				Treble Low Water[3,5] Treble Wet[3,5]				B380[3,5]			X2B[3] GRSX2B[3,4] CS12[3,5]		
Salt Water	2546[3,4]	930	254SS[3] L067[3,4]		SS15[3,4]				34007 3407[4] 31014[3] 3366[3,4] 3406[4] 3406B[4] 34009	9034[3,4]	CS51[3,4] CS41[3,4]		9752[3,4]

Turrall	Daiichi	Dai-Riki	Eagle Claw	Gaelic Supreme	Gamakatsu	Hayabusa	Heritage	Kamasan	Mustad	Orvis	Partridge	Varivas	VMC
Sedge	1150[3] 1155[3,4]			Caddis[3,4]		371			81001BR		K2B		
Sproat	1550	070[3]	L057		P-10[3] ***R10-BN***[3,4] S10[3,4] ***S10-B***[3,4]		Trout Down Eye		3906 3399 3399A 3399N[4]		G3A J1A[3] CS7MW[3,4]		8526
Sproat Double				Double Low Water[3,4] Double Wet[3,4]				B280[4]			R2A[3] Q1[3,4] GRSQ2[3,4]		9410[3]
Stronghold						753[3]	Stronghold					973[3]	
Tube Fly Treble								B990[4]	3551[3,4] 35647[4]		X3 BL[3] X3 SI[3,4]		
Up-eye Dry Fly							Trout Up Eye						

Hooks that are ***bold italicized*** are barbless.
[1] Turrall considers this hook to be similar. The brand at the top of the column may or may not concur.
[2] The brand at the top of the column considers the Turrall hook to the left to be similar. Turrall may or may not concur.
[3] Similar in all aspects except the bend. See the corresponding characteristic tables.
[4] Similar in all aspects except the finish. See the corresponding characteristic tables.
[5] Similar in all aspects except the point position. See the corresponding characteristic tables.

TURRALL Characteristics

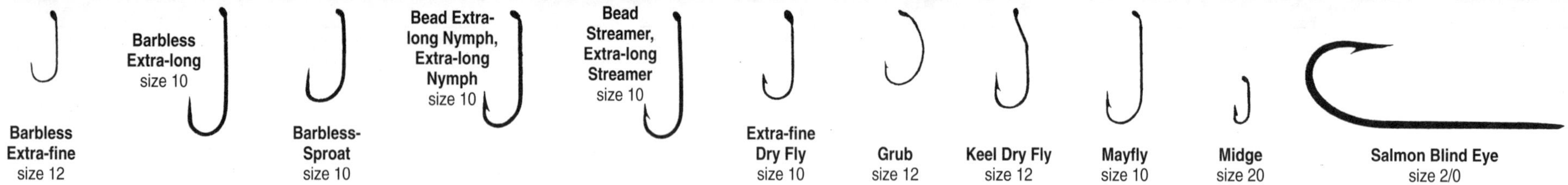

Hook	Use	Wire	Shank	Eye	Bend	Finish	Barb	Point	Common Name	Remarks	Size
Barbless Extra-fine	Dry	4XF	S	TD	P	Bron	No Barb	Chem	Barbless-Dry Fly	Forged.	12-18
Barbless Extra-long	Streamer	S	3XL	TD	P	Bron	No Barb	Chem	Barbless-Streamer	Long-bodied nymphs. Forged.	8, 10
Barbless-Sproat	Nymph	S	S	TD	Spr	Bron	No Barb	Chem	Barbless-Sproat (Wet Fly)	Wet flies. Forged.	10-14
Bead Extra-long Nymph	Nymph	S	1XL	TD	P-WG	Bron	Mini	Chem	Bead-Extra-long Nymph	2x WG to fit bead heads. Available with bead attached. Forged.	10-14
Bead Standard Nymph	Nymph	S	S	TD	P-WG	Bron	Mini	Chem	Bead-Standard Nymph	2x WG to fit bead heads. Available with bead attached. Forged.	10-14
Bead Streamer	Streamer	S	3XL	TD	P-WG	Bron	Mini	Chem	Bead-Streamer	2x WG to fit bead heads. Available with bead attached. Forged.	8
Extra-fine Dry Fly	Dry	4XF	S	TD	P	Bron	Mini	Chem	Extra Fine-Dry Fly	Forged.	8-18
Extra-long Nymph	Nymph	S	1XL	TD	P	Bron	Mini	Chem	Extra Long-Nymph	Forged.	8-12
Extra-long Streamer	Streamer	S	3XL	TD	P	Bron	Mini	Chem	Extra Long-Streamer	Long-bodied nymphs. Forged.	2-12
Grub	Scud/Shrimp	S	S-C	TD	C-WG	Bron	Mini	Chem	Grub	Grubs and larvae. Forged.	12-16
Keel Dry Fly	Dry	2XF	1XL-BD	S	Spr	Bron	Mini	Chem	Keel-Dry Fly	Mainly for "upside-down" dry flies. Forged.	12-16
Mayfly	Dry	4XF	1XL	TD	P	Bron	Mini	Chem	Mayfly	Forged.	8-12
Midge	Midge/Pupae	3XF	S	TD	P	Bron	Mini	Chem	Midge	Small dry flies. Forged.	20-24
Salmon Blind Eye	Display	S	S	No	Lim	Blk	Mini		Salmon Blind Eye	Forged.	4/0-2/0

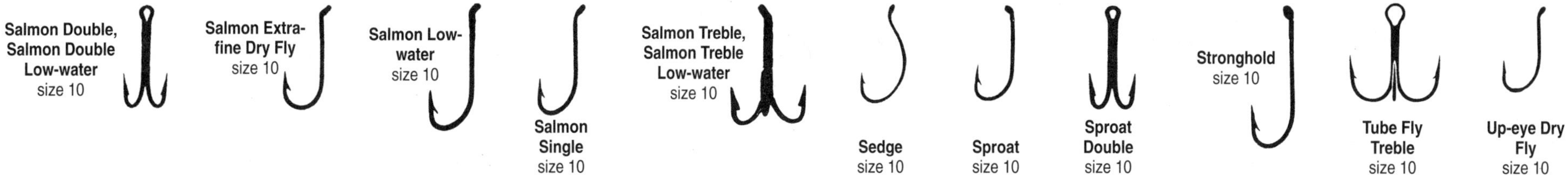

Hook	Use	Wire	Shank	Eye	Bend	Finish	Barb	Point	Common Name	Remarks	Size
Salmon Double	Double	S	S	TU-L	Lim-Dbl	Blk	Mini	Chem	Salmon Double	Forged.	2/0-12
Salmon Double Low-water	Double	S	2XL	TU-L	Lim-Dbl	Blk	Mini	Chem	Salmon Low-water-Double	Forged.	2-12
Salmon Extra-fine Dry Fly	Steelhead/ Salmon	2XF	S	TU-L	Lim	Blk	Mini	Chem	Salmon Extra-fine Dry	Forged.	2-10
Salmon Low-water	Steelhead/ Salmon	S	2XL	TU-L	Lim	Blk	Mini	Chem	Salmon Low-water	Forged.	1-12
Salmon Single	Steelhead/ Salmon	S	S	TU-L	Lim	Blk	Mini	Chem	Salmon Single	Forged.	4/0-12
Salmon Treble	Treble	S	2XL	TU-L	Lim-Treb	Blk	Mini	Out-C	Salmon Treble	Forged.	2-10
Salmon Treble Low-water	Treble	S	S	TU-L	Lim-Treb	Blk	Mini	Out-C	Salmon Low-water Treble	Forged.	2-10
Salt Water	Saltwater	S	S	S	Osh	SS	Mini		Salt Water	Forged.	3/0-6
Sedge	Caddis	S	S-C	TU	C-WG	Bron	Mini	Chem	Sedge	Scuds and shrimp. Forged.	8-16
Sproat	Nymph	S	S	TD	Spr	Bron	Mini		Sproat Nymph/Wet Fly	Wet flies. Forged.	2-16
Sproat Double	Double	S	S	TD-L	Spr-Dbl	Bron	Mini	Chem	Sproat Double	Nymphs & wet flies. Forged.	4-16
Stronghold	Streamer	3XH	3XL	TD	Spr	Bron	Mini	Chem	Stronghold X-Strong Streamer	Long-bodied nymphs. Forged.	4-10
Tube Fly-Treble	Treble	S	S	S	P-Treb	Blk	Mini	Chem	Tube Fly Trebles	Tube flies. Forged.	2-12
Up-eye Dry Fly	Dry	3XF	S	TU	Spr-R	Bron	Mini	Chem	Up-Eye Dry Fly	Forged.	8-18

Varivas	Daiichi	Dai-Riki	Eagle Claw	Gaelic Supreme	Gamakatsu	Hayabusa	Heritage	Kamasan	Mustad	Orvis	Partridge	TMC	Turrall	VMC
98i									3399D[3,4]		L2A[3,4] GRS2A[3,4] ***CS27GRS***[3,4]	***902BL***		V7060[3]
98r	1550[3,4]		L057[3,4]			373[3]			3399[3,4] 3399A[3,4] 3399N[3,4] 3906[3,4] 7948A[3,4]		J1A[4] CS7MW[3] G3A[3,4]		Sproat[3,4] ***Barbless Sproat***[3,4] Bead Standard Nymph[3,4]	8526[3,4]
971														
972														
973						753[3]	Stronghold[3]						Stronghold[3]	
974	2441[3]			Single Low Water[3]				B190[3]		1645[3]	M[3]	7999[3]		
977		075[3,4]			L10-2H[3,4]						SH1[4]	3769[3,4]		
980	1170 1180 ***1190*** 1182[4]	305				372 373[3] ***378*** 386		***B402*** K40 B401[3]	7948A	1509[3] ***1877***[5] 4864[3]	CS7MW[3,4]		Bead Standard Nymph[3]	9060
981						752[3]			94833[3]					9288[3,4]
988					S-13S-M[3]				94859[3]			501[3,4]		

Hooks that are ***bold italicized*** are barbless.
[1]Varivas considers this hook to be similar. The brand at the top of the column may or may not concur.
[2] The brand at the top of the column considers the Varivas hook to the left to be similar. Varivas may or may not concur.
[3] Similar in all aspects except the bend. See the corresponding characteristic tables.
[4] Similar in all aspects except the finish. See the corresponding characteristic tables.
[5] Similar in all aspects except the point position. See the corresponding characteristic tables.
[6] Weight and length varies. Eye, bend, finish and shape are similar unless otherwise noted.

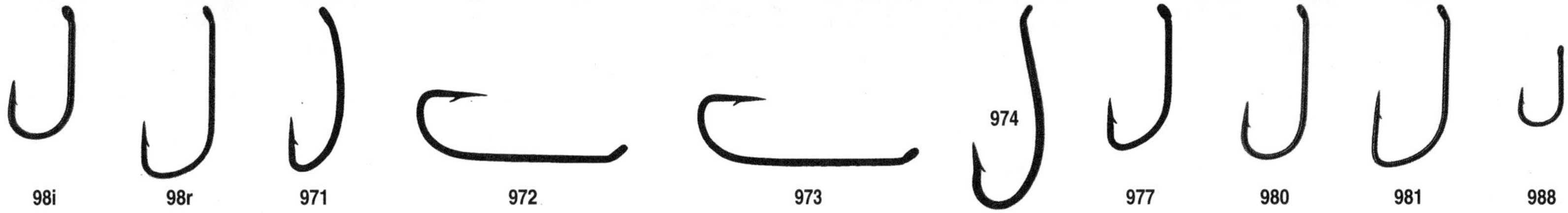

Hook	Use	Wire	Shank	Eye	Bend	Finish	Barb	Point	Common Name	Remarks	Size
98i	Emerger	2XF	S	TD-M	P-M	Blk	Mini	Chem	Area Indicator	Eye partially TD. WG P bend.	12-20
98r	Nymph	S	S	TD	Lim-M	Blk	Mini	Chem	Area Retrieve	Justice Lim bend creates a wider gape.	8-16
971	Nymph	1XH	1XL-C	TD-M	Lim-M	Bron	Mini	Chem	Nymph		8-20
972	Streamer	3XH	2XL	TD	Lim	Bron	Mini	Chem	Streamer & Nymph	Nymphs.	4-12
973	Streamer	3XH	3XL	TD	Lim	Bron	Mini	Chem	Streamer & Nymph	Nymphs.	4-12
974	Steelhead/ Salmon	1XH	S	TU-L	Spr-M	Blk	Mini	Chem	Up-eye Salmon		2-10
977	Wet	2XH	S	TD	Lim	Blk	Mini	Chem	Wet	Nymphs.	6-14
980	Dry	S	S	TD	P	Bron	Mini	Chem	Basic Dry	Wet flies.	10-22
981	Dry	2XF	1XS	TD	Lim-M	Blk	Mini	Chem	IWI Dry	WG Lim bend.	10-22
988	Midge/Pupae	2XF	1XS	S	P-WG	Blk	Mini	Chem	Midge	WG P bend.	18-24

VMC	Daiichi	Dai-Riki	Eagle Claw	Gaelic Supreme	Gamakatsu	Hayabusa	Kamasan	Mustad	Orvis	Partridge	TMC	Turrall	Varivas
7250NI	3111[3,4]		L254N L253 254N 254[4] 254SS[3,4] L254BK[4] L067[3,4]		SC15[3,4]					Z12[3,4] Z1[3,4] Z2[3,4] ***Z1Y***[3,4] Z10[4] GRS 1Z[3,4] MM3ST[3,4]			
7255	X452[4]		250[3,4]				K93[3,4]		XPoint[4]				
7299	2550 2552 2553 2556[4] 4250 4253 4255[4]												
7299SS	2556 2550[4] 2552[4] 2553[4] 4250[4] 4253[4] 4255[4]												
7399	2550 2552 2553 2556[4] 4250 4253 4255[4]												
8410	3111[3,4]		L254N[4] L253[4] 254N[4] 254[4] 254SS[3,4] L254BK[4] L067[3,4]		SC15[3,4]					Z12[3,4] Z1[3,4] Z2[3,4] ***Z1Y***[3,4] Z10[3,4] GRS 1Z[3,4] MM3ST[3,4]			

Hooks that are ***bold italicized*** are barbless.
[1]VMC considers this hook to be similar. The brand at the top of the column may or may not concur.
[2]The brand at the top of the column considers the VMC hook to the left to be similar. VMC may or may not concur.
[3]Similar in all aspects except for the bend. See the corresponding characteristic tables.
[4]Similar in all aspects except the finish. See the corresponding characteristic tables.
[5]Similar in all aspects except the point position. See the corresponding characteristic tables.

VMC	Daiichi	Dai-Riki	Eagle Claw	Gaelic Supreme	Gamakatsu	Hayabusa	Kamasan	Mustad	Orvis	Partridge	TMC	Turrall	Varivas
8526	1530[2] 1550	305[3]	L057		S10[3,4]	373		3906[1] 3399 3399A 3399N[4] 7948A[3]	1641[2]	G3A J1A[3] CS7MW[3,4]		Sproat ***Barbless Sproat*** Bead Standard Nymph[3]	98r[3,4]
8527	1550[2]	300[3]	L057				B170	3906B[1] AC3906B 7957B[3] 3399A		G3A		Extra-long Nymph[3] Bead Extra-long Nymph[3]	
8913													
9060	1170 1180 1182[4] ***1190***	305			P-10 S10[4] ***S10-B***[4]	372 373[3] ***378*** 386	B401[3] ***B402*** K40		***1877***[5] 4864[3] 1509[3]				980
9148													
9255	1750[2,3]						B940[3,4]			D3ST[3]			
9279	1710[2,3]		L063		P10-2L1H[3]		B200[3]	9671[1]	1524[2]	D4A			
9280	1170[2,3] 1100[3]		***L061B***[2,3] L059[3] L060[3]		S10-3F[3,4]	572[3]	B400[3]	94840[3] AC94840[3] ***94845***[3] 80000BR[3] ***80010BR***[3]	1523[2,3]	L2A[3] GRS2A[3,4] L3A[3] ***L3AY***[3] GRS3A[3,4]	5210[3] 100 ***100BL*** 102Y[3,4] ***103BL***[4] ***900BL***[3,4,5] 146Z[3,4]		
9281							K18[3]	94842[1,3]					
9282						751[3]	B405[3]	94838[3]		***CS20***[3,4,5]			981[3]
9283		700					B800[3] K80	79580[1] 33960[3] 3498AD[3,4]	1511[2,3] 8808	CS11/1[3,4] CS11/2[3,4] CS11GRS[3,4] ***CS2002***[5]			
9288						752[3]		94833[3]			5230		981[3,4]
9289	1330[2]												

VMC	Daiichi	Dai-Riki	Eagle Claw	Gaelic Supreme	Gamakatsu	Hayabusa	Kamasan	Mustad	Orvis	Partridge	TMC	Turrall	Varivas
9365													
9388							B405[3]	94838[3]		***CS20***[3,4,5]			
9410				Double Low Water[3,4] Double Wet[3,4]			B280[3,4]			Q1[3,4] GRSQ2[3,4]		Salmon Double Low-water[3,4] Salmon Double[3] Sproat Double[3]	
9752	2546[3,4]				SS15[3,4]			31014[3,4] 34007[3,4] 3407[3,4] 3366[3,4] 3406[3,4] 3406B[3,4] 34009[3,4]	9034[3,4]	CS51[3,4] CS41[3,4]		Salt Water[3,4]	
9902			L056[3]		C12[3]							Grub[3]	
V7060	1180[1] 1182[1,4]	305[1]						94840[1] 3399D[3]		L2A[1,3] ***CS27GRS***[3,4] GRS2A[3,4]	5210[1] ***902BL***[3,4]		98i[3]
V7061	1100[1]		L059 L060[3] ***L061B***		S10-3F[4]	572[3]	B400[3]	94840 AC94840 ***94845*** 94834[4] 80000BR ***80010BR***	1523	L3A[1,3] ***L3AY***[3] GRS3A[3,4]	5210 100[1,3] ***100BL***[3] 102Y[3,4] ***103BL***[3,4] ***900BL***[3,4,5] 146Z[4]		
V7062	1280[1]	730[1]				552		94831[1]		H1A[1,3]	5212[1]		
V7070	1720[1,3]	710[1,3]						AC9672[1,3]			5263[1,3] ***5263BL***[3]		
V7071	1560[1]							3906B[1]		SH2[1,3,4]	3761[1] ***3761BL***		

Hooks that are ***bold italicized*** are barbless.
[1]VMC considers this hook to be similar. The brand at the top of the column may or may not concur.
[2]The brand at the top of the column considers the VMC hook to the left to be similar. VMC may or may not concur.
[3]Similar in all aspects except for the bend. See the corresponding characteristic tables.
[4]Similar in all aspects except the finish. See the corresponding characteristic tables.
[5]Similar in all aspects except the point position. See the corresponding characteristic tables.

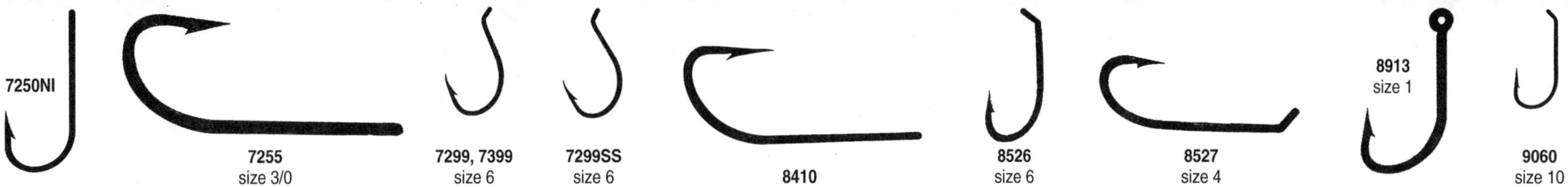

Hook	Use	Wire	Shank	Eye	Bend	Finish	Barb	Point	Common Name	Remarks	Size
7250NI	Bass	S	S	S	Aber	Nick				Tube flies.	5/0-1/0
7255	Saltwater	S	1XL	S	Osh	SS	Mini	Chem		Forged.	7/0-6
7299	Salmon	S	1XS	TU	Beak-R	(Rmks)				Same as 7399. Available in bron, nick, red & sil. Octopus style hook. Forged.	11/0-6
7299SS	Salmon	S	1XS	TU	Beak-R	SS				Octopus style hook. Forged.	9/0-6
7399	Salmon	S	1XS	TU	Beak-R	(Rmks)	No Barb		Octopus	Same as 7299. Available in bron, nick, red & sil. Forged.	3/0-4
8410	Saltwater	S	S	S	Osh	(Rmks)			Crystal	Bass. Available in sil & gold. Forged.	5/0-2
8526	Nymph	S	S	TD	Spr	Bron				Forged.	6-16
8527	Nymph	S	1XL	TD	Spr	Bron				Forged.	4-16
8913	Wet	1XF	1XS	F	Osh	(Rmks)	Mini	Chem		Available in bron & nick.	1-20
9060	Dry	S	S	TD	P	Bron				Forged.	8-22

VMC Characteristics

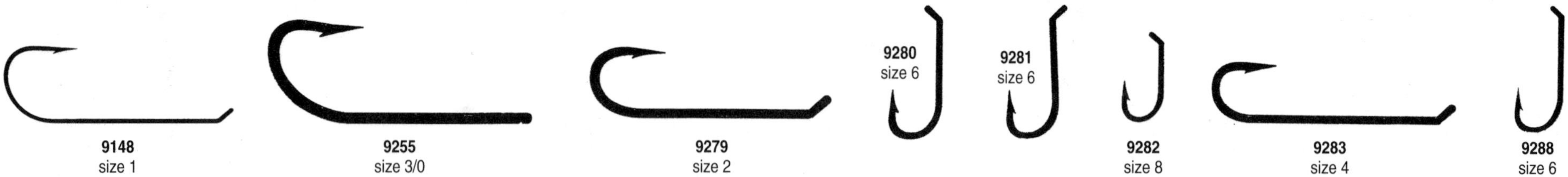

Hook	Use	Wire	Shank	Eye	Bend	Finish	Barb	Point	Common Name	Remarks	Size
9148	Bass	S	S	TD	Aber	(Rmks)				Available in bron & gold.	4/0-1
9255	Streamer	1XH	4XL	S	Osh	(Rmks)				Available in bron, nick & sil. Forged.	9/0-8
9279	Nymph	1XH	2XL	TD	P-M	Bron				Forged.	2-18
9280	Dry	1XF	S	TD	P-M	Bron				Forged.	6-20
9281	Dry	1XF	S	TU	P-M	Bron				Forged.	8-18
9282	Dry	1XF	1XS	TD	P-R	Bron				Forged.	8-18
9283	Streamer	S	4XL	TD	P	Bron				Forged.	2-18
9288	Dry	2XF	1XS	TD	P	Bron				Forged.	6-20

Hook	Use	Wire	Shank	Eye	Bend	Finish	Barb	Point	Common Name	Remarks	Size
9289	Dry	2XF	1XS	TU	P	Bron				Forged.	6-20
9365	Egg	S	2XS	TU	C	Gold			Salmon Egg	Forged.	2-14
9388	Dry	1XF	1XS	TD	P	(Rmks)	No Barb	Chem		Available in bron & gold.	6-20
9410	Double	S	1XL	TU-L	Lim-Dbl	Bron			Double Fly		4-8
9752	Saltwater	S	S	S	WG-R	Sil			Sea Kirby		1-20
9902	Nymph	S	S-C	TD	C-WG-R	(Rmks)	Mini	Chem		Scud/shrimp & larvae. Available in bron & nick.	8-14
V7060	Dry	2XF	S	TD	P	Bron	Mini	Chem	Needle Cone Standard Dry Fly		8-22
V7061	Dry	1XF	S	TD	P	Bron	Mini	Chem	Needle Cone Dry Fly		6-24
V7062	Dry	1XF	2XL	TD	P	Bron	Mini	Chem	Needle Cone Dry Fly		6-16
V7070	Nymph	2XH	3XL	TD	Spr	Bron	Mini	Chem	Needle Cone Nymph Wet	Wet flies & streamers.	4-18
V7071	Nymph	2XH	1XL	TD	Spr	Bron	Mini	Chem	Needle Cone Nymph Wet	Wet flies & streamers.	4-18

BARBLESS HOOKS

Maker	Hook	Chief Use	Size
Daiichi	1190	Dry	8-24
Daiichi	1250	Scud/Shrimp	12-18
Daiichi	1251	Scud/Shrimp	12-18
Eagle Claw	L061B	Dry	4-18
Eagle Claw	L064B	Steelhead/Salmon	1-8
Gamakatsu	R10-BN	Dry	10-20
Gamakatsu	S10-B	Dry	6-16
Hayabusa	378	Dry	8-24
Hayabusa	379	Dry	16-24
Hayabusa	381	Nymph	10-18
Hayabusa	382	Streamer	8-18
Hayabusa	388	Caddis	8-20
Kamasan	B402	Dry	14-20
Mustad	80010BR	Dry	10-14
Mustad	80060BR	Nymph	10-14
Mustad	80260BR	Scud/Shrimp	12-16
Mustad	94845	Dry	8-20
Orvis	1877	Dry	8-24
Partridge	CS20	Dry	10-18
Partridge	CS2001	Dry	14-16
Partridge	CS2002	Streamer	8-10
Partridge	CS2003	Nymph	12-14
Partridge	CS27GRS	Dry	10-18
Partridge	CS28GRS	Nymph	8-14
Partridge	CS29GRS	Streamer	2-8
Partridge	E3AY	Dry	10-18
Partridge	L3AY	Dry	10-16
Partridge	Z1Y	Carp	1-10
Partridge	Z31	Carp	4-10
TMC	100BL	Dry	10-22
TMC	103BL	Dry	11-19
TMC	109BL	Dry	7-19
TMC	200RBL	Nymph	4-22

BARBLESS HOOKS (continued)

Maker	Hook	Chief Use	Size
TMC	205BL	Scud/Shrimp	8-20
TMC	206BL	Nymph	6-20
TMC	2487BL	Scud/Shrimp	12-22
TMC	3761BL	Nymph	10-18
TMC	5263BL	Nymph	4-12
TMC	900BL	Dry	8-24
TMC	902BL	Dry	12-20
TMC	905BL	Dry	2-10
TMC	947BL	Nymph	4-20
Turrall	Barbless Extra-fine	Dry	12-18
Turrall	Barbless Extra-long	Streamer	8, 10
Turrall	Barbless-Sproat	Nymph	10-14
VMC	7399	Salmon	3/0-4
VMC	9388	Dry	6-20

BLIND-EYED HOOKS

Maker	Hook	Bend
Daiichi	2091	Lim-M
Gaelic Supreme	F-F Light Blind Eye	DLim
Gaelic Supreme	F-F Spey Style Blind Eye	DLim
Gaelic Supreme	F-F Dee Style Blind Eye	DLim
Gaelic Supreme	F-F Blind Eye	Dlim
Heritage	Pryce-Tannant Pattern	DLim
Heritage	Harrison-Bartleet Pattern	DLim
Partridge	CS10/3	Dub
Partridge	CS6	Lim-M
Turrall	Salmon Blind Eye	Lim

OVER-SIZED EYED HOOKS

Maker	Hook	Use	Eye	Size
Daiichi	1100	Dry	TD-O	16-24
Daiichi	1110	Dry	S-O	12-26
Gamakatsu	Trey Combs Big Game	Saltwater	S-O	10/0-2
Hayabusa	377	Dry	S-O	12-24
Hayabusa	379	Dry	S-O	16-24
Kamasan	B990	Treble	S-O	2-14
Mustad	34009	Saltwater	S-O	5/0-4
Orvis	4641	Dry	S-O	16-24
Orvis	4864	Dry	TD-O	16-24

FLAT-EYED HOOKS

Maker	Hook	Use	Eye	Bend
Daiichi	1850	Streamer	F	P
Partridge	K10	Dry	F-L	P-M
Partridge	K17	Dry	F-M	P
VMC	8913	Wet	F	Osh

BASS HOOKS

Maker	Hook	Bend	Finish	Size
Dai-Riki	930	Osh	SS	3/0-1/0, 2-8
Daiichi	2451	Osh	Blk	4/0-8
Daiichi	2720	WG	Bron	5/0-2
Daiichi	2722	WG	Nick	5/0-2
Daiichi	X452	Osh	Sil	4/0-6
Eagle Claw	066	Lim-M-R	Corr-Sil	7/0-6
Eagle Claw	066N	Lim-M-R	Nick	5/0-8
Eagle Claw	066SS	Lim-M-R	SS	6/0-4
Eagle Claw	178	Spr-M	Gold	2-10
Eagle Claw	254SS	Osh-R	SS	9/0-2
Kamasan	B940	Aber	Blk	6/0-6
Mustad	3366	Spr	Bron	5/0-14
Mustad	3498AD	Osh	Corr-Sil	5/0-4, 10, 8
Mustad	37187	WG	Bron	1/0-10
Mustad	79666S	P	SS	1/0-6
Mustad	7970	Lim	Bron	1-8
Mustad	90233S	Spr-WG	SS	2/0

BASS HOOKS (continued)

Maker	Hook	Bend	Finish	Size
Orvis	XPoint	Osh	Sil	4/0-8
Partridge	Z10	Aber	Blk	6/0-8
TMC	8089	Lim-M	Bron	2, 6, 10
TMC	8089NP	Lim-M	Nick	2, 6, 10
TMC	9394	Osh	Nick	2-10
TMC	9395	Osh	Bron	2-10
VMC	8410	Osh	Sil & Gold	5/0-2

POPPER HOOKS

Maker	Hook	Finish	Kinks	Size
Eagle Claw	L200	Bron	Double	1/0-12
Mustad	33900	Bron	Single	1/0-12
Mustad	33903	Bron	Single	1/0-14
Mustad	34015D	Corr-Sil	Double	1/0
Mustad	90233S	SS	Single	2/0
TMC	511S	SS	Single	2/0-1/0, 2,6

BENT-DOWN SHANK HOOKS

Maker	Hook	Use	Eye	Size
Dai-Riki	700-B	Nymph	TD	2-16
Daiichi	1730	Nymph	TD	4-14
Daiichi	1870	Nymph	TU	4-14
Gamakatsu	C11-5L2H	Nymph	TD	6-12
Kamasan	B810	Nymph	TD	6-14
Mustad	81002BR	Nymph	TD	6-14
Partridge	Z13	Carp	S	2-8
Turrall	Keel Dry Fly	Dry	S	12-16

BENT-UP SHANK HOOKS

Maker	Hook	Use	Eye	Size
Daiichi	1770	Nymph	S	6-16
Daiichi	2170	Egg	S	2-12
Daiichi	2171	Egg	S	2-12
Daiichi	2174	Egg	S	2-12
Hayabusa	380	Nymph	S	8-14
Mustad	34005	Saltwater	S	5/0-1
Mustad	80150BR	Nymph	S	8-14
Orvis	1512	Nymph	S	6-16
Partridge	H1B/CA	Nymph	TU-M	2-18
Partridge	K3A	Dry	TD	10-18
Partridge	K6ST	Nymph	S	8-14
Partridge	K7	Dry	TU	10, 12
TMC	400T	Nymph	S	8-14
TMC	411S	Saltwater	S	3/0, 1/0, 4-8
TMC	947BL	Nymph	TD	4-20

UNIQUE HOOKS

Maker	Hook	Use	Remarks
Partridge	H3ST	Nymph/Draper	Shank is split lengthwise for wide, flat bodies.
Partridge	K10	Dry	Shank is looped over top and extends past bend then curves upward.
Partridge	K16B	Dry	Shank has distinct kink at thorax. Designed by Jens Pilgaard.
Partridge	X7	Treble	Shank has extended spike to fit into tubes of tube flies.

SALTWATER HOOKS

Maker	Hook	Maker	Hook
Dai-Riki	930	Orvis	9034
Daiichi	2546	Orvis	XPoint
Daiichi	X452	Owner	5106
Eagle Claw	066	Owner	5111
Eagle Claw	066N	Owner	5170
Eagle Claw	066SS	Owner	5320
Eagle Claw	178	Partridge	CS51
Eagle Claw	250	Partridge	CS52
Eagle Claw	250H	Partridge	CS53
Eagle Claw	254	Partridge	CS54
Eagle Claw	254N	Partridge	GRS S3
Eagle Claw	254SS	Partridge	GRS S4
Eagle Claw	L054SS	Partridge	S3
Eagle Claw	L067	Partridge	S4
Eagle Claw	L072BK	Partridge	Z10
Eagle Claw	L253	Partridge	Z8
Eagle Claw	L254BK	TMC	411S
Eagle Claw	L254N	TMC	800S
Eagle Claw	L256	TMC	811S
Eagle Claw	L7254	TMC	911S
Eagle Claw	NT156	Trey Combs	Big Game
Gamakatsu	SC15	Turrall	Salt Water
Gamakatsu	SL11-3H	VMC	7255
Gamakatsu	SP11-3L3H	VMC	8410
Gamakatsu	SS15	VMC	9752
Kamasan	B940		
Mustad	31014		
Mustad	34005		
Mustad	34007		
Mustad	34009		
Mustad	34011		
Mustad	34039SS		
Mustad	3407		
Mustad	3407SS		
Mustad	3408B		
Mustad	3498AD		
Mustad	37160S		
Mustad	77660SS		
Mustad	79666S		
Mustad	90233S		

TREBLE HOOKS

Maker	Hook	Size
Kamasan	B380	4-16
Kamasan	B990	2-14
Gaelic Supreme	Treble Low Water	4-12
Gaelic Supreme	Treble Wet	4-12
Mustad	3551	5/0-20
Mustad	35647	3/0-14
Mustad	80550BL	1/0-4
Partridge	CS12	2-12
Partridge	CS18	12-16
Partridge	CS8ST	12-16
Partridge	CS9 BL	2-10
Partridge	CS9 SI	2-10
Partridge	GRS X1	3/0-10
Partridge	GRS X2B	6-10
Partridge	GRS X8	4-10
Partridge	X1 BL	3/0-16
Partridge	X1 BR	3/0-16
Partridge	X1 SI	3/0-16
Partridge	X2B	2-16
Partridge	X3 BL	2-16
Partridge	X3 SI	4-12
Partridge	X7	4-12
Turrall	Salmon-Treble	2-10
Turrall	Salmon-Treble Low-water	2-10
Turrall	Tube Fly-Treble	2-12

DOUBLE HOOKS

Maker	Hook	Size
Daiichi	7131	4-12
Gaelic Supreme	Double Low Water	2/0-12
Gaelic Supreme	Double Wet	2/0-12
Kamasan	B270	6-14
Kamasan	B280	4-12
Mustad	80525BL	2-10
Mustad	80526BL	2-10
Orvis	0528	4-12
Partridge	CS10/4	4-8
Partridge	GRSQ2	6-10
Partridge	O2	2-16
Partridge	P	2-12
Partridge	Q1	3/0-12
Partridge	R1A	2-12
Partridge	R3HF	6-14
Turrall	Salmon Double	2/0-12
Turrall	Salmon Double Low-water	2-12
Turrall	Sproat Double	4-16
VMC	9410	4-8

SINGLE HOOKS FOR TUBE FLIES

The ideal single hook for a tube fly is a straight open eye with a short or standard shank.

Maker	Hook	Wire	Shank	Eye	Bend	Finish
Daiichi	1640	S	2XS	S	P-R	Bron
Daiichi	2451	S	1XS	S	Osh	Blk
Daiichi	2546	S	S	S	Osh	SS
Daiichi	X452	S	1XS	S	Osh	Sil
Eagle Claw	254	S	S	S	Osh	Corr-Sil
Eagle Claw	254N	S	S	S	Osh	Nick
Eagle Claw	L253	S	S	S	Osh	Nick
Eagle Claw	L254BK	S	S	S	Osh	Blk
Eagle Claw	L254N	S	S	S	Osh	Nick
Eagle Claw	L7254	1XH	S	S	Osh	Rmks-Nick
Mustad	34007	S	S	S	Osh	SS
Mustad	3407	S	S	S	Osh	Tin
Mustad	77660SS	1XH	1XS	S	Osh	SS
Orvis	XPoint	S	1XL	S	Osh	Sil
Partridge	MM3ST	S	S	S	Spr-M	Blk
TMC	105	2XH	7XS	S	P-M-R	Bron
TMC	800S	1XH	S	S	Osh	SS
TMC	811S	1XH	S	S	Osh	SS
VMC	7250NI	S	S	S	Aber	Nick

HOOKS WITH COLOR

Hooks that are *not* bronzed, blackened (any method), or variations of silver or tin.

Maker	Hook	Use	Finish	Size
Daiichi	1155	Scud/Shrimp	Gold	8-18
Daiichi	1273	Nymph	Red	8-22
Daiichi	2055	Steelhead/Salmon	Gold	3/0-7
Daiichi	2059	Steelhead/Salmon	Blue	3/0-7
Daiichi	2065	Steelhead/Salmon	Gold	3
Daiichi	2135	Steelhead/Salmon	Gold	2-8
Daiichi	2139	Steelhead/Salmon	Blue	2-8
Daiichi	2174	Egg	Grn	2-12
Daiichi	2553	Salmon	Red	6-2
Daiichi	4253	Egg	Red	4-14
Daiichi	4255	Egg	Gold	4-14
Eagle Claw	178	Saltwater	Gold	2-10
Eagle Claw	L1197G	Steelhead/Salmon	Gold	1-8
Gaelic Supreme	Caddis	Scud/Shrimp	Olive	8-16
Gaelic Supreme	Grub Hook	Scud/Shrimp	Olive	8-16
Gaelic Supreme	Nymph Hook	Nymph	Olive	8-16
Gaelic Supreme	Wet Fly Hook	Wet	Olive	6-16
Gamakatsu	T10-6H	Steelhead/Salmon	Rmks	2/0-8
Kamasan	B100G	Scud/Shrimp	Gold	10-16
Mustad	3399N	Nymph	Gold	10/0-2
Mustad	37162	Scud/Shrimp	Gold	4/0-10
Mustad	94834	Hook Discontinued	Gold	N/A
Mustad	9674A	Streamer	Gold	8-14
Owner	5111	Saltwater	Rmks	8/0-8
Partridge	CS5	Streamer	Blue	4-8
Partridge	H1B/CA	Nymph	Rmks	2-18
Partridge	Z31	Carp	Blue	4-10
TMC	2487G	Scud/Shrimp	Gold	10-20
VMC	7299	Salmon	Rmks	11/0-6
VMC	7399	Salmon	Rmks	3/0-4
VMC	8410	Saltwater	Rmks	5/0-2
VMC	9148	Bass	Rmks	4/0-1
VMC	9365	Egg	Gold	2-14
VMC	9388	Dry	Rmks	6-20

STAINLESS STEEL HOOKS

Maker	Hook
Dai-Riki	930
Daiichi	2546
Daiichi	2556
Eagle Claw	066SS
Eagle Claw	254SS
Eagle Claw	L054SS
Mustad	31014
Mustad	34005
Mustad	34007
Mustad	34009
Mustad	34011
Mustad	34039SS
Mustad	37160S
Mustad	77660SS
Mustad	79666S
Mustad	90233S
Orvis	9034
Partridge	CS11/2
Partridge	CS11/3
Partridge	CS52
Partridge	CS53
Partridge	Z8
Tiemco	411S
Tiemco	511S
Tiemco	800S
Tiemco	811S
Tiemco	911S
Turrall	Salt Water
VMC	7255
VMC	7299SS

MANUFACTURER CONTACT LIST

Manufacturers request that retail customers contact local fly shops for hooks before ordering from this list.

Daiichi
Angler Sport Group
6619 Oak Orchard Rd.
Elba, NY 14058
(716) 757-9958
FAX Orders: (877) 757-9066
E-mail Orders:
orders@anglersportgroup.com

Dai-Riki
Dan Bailey's Fly Shop
PO Box 1019
Livingston, MT 59047
(800) 356-4052
(406) 222-1673
FAX: (406) 222-8450
Web: www.dan-bailey.com

Eagle Claw
Wright & McGill Co.(Wholesale)
4245 E. 46th Ave.
PO Box 16011
Denver, CO 80216-0011
(800) 432-HOOK (4665)
(wholesale orders only)
(303) 321-1481
FAX: (303) 321-4750
Email: custsvc@eagleclaw.com
Web: www.eagleclaw.com

Gaelic Supreme
Belvoirdale
PO Box 176
Wyncote, PA 19095-0176
(215) 886-7211
FAX: (215) 886-1804
e-mail:
belvoirdale@belvoirdale.com
Web: www.belvoirdale.com

Gamakatsu
Gamakatsu, USA
PO Box 1797
Tacoma, WA 98401
(253) 922-8373
Wholesale: 1-800-223-9383 (Sales ext. 300)
FAX: (253) 922-8447
Web: www.gamakatsu.com

Hayabusa
Hayabusa USA, Inc.
2585 White Dove Lane
Chino Hills, CA 91709
(909) 628-2948
FAX: (909) 628-0849
Web: www.hayabusa.co.jp

Hayabusa Fishing Hooks Co., Ltd.
Watase Yokawa-Cho,
Mino-Gun, Hyogo
673-11, Japan
(0794)73-0212
FAX: (0794)73-200

Heritage
Castle Arms
PO Box 30070
Springfield, MA 01103
(413) 567-8268
FAX: (413) 731-1292
Wholesale Orders: 1-800-525-4866

Kamasan
Kamatsu Kamasan, Ltd.
(Wholesale)
Unit 1 Sanford Lane Industrial Estate
Kennington, Oxford
OX1 5RW, U.K.
(01856) 326655
FAX: (01865) 326650

Mustad
O. Mustad & Son (U. S. A.) Inc.
(Wholesale)
PO Box 838
247-253 Grant Ave.
Auburn, NY 13201
(800) 453-4540
(315) 253-2793
FAX: (315) 253-0157
Web: www.mustad.no

Orvis
Orvis Company
PO Box 798
Historic Route 7A
Manchester, VT 05254-0798
(800) 548-9548
(802) 362-3622
Web: www.orvis.com

Owner
Owner American Corporation
3199-B Airport Loop Dr.
Costa Mesa, CA 92626
(714) 668-9011
FAX: (714) 668-9133

Owner Hooks Co., Ltd
120 Tomita Nishiwaki-City
Hyogo Pref. 677-0018 Japan
81-(0) 795-22-1433
FAX: 81-(0) 795-22-6003

Partridge
Partridge of Redditch LTD.
Mount Pleasant
Redditch, Worcestershire,
England B97 4JE
011 44 1527 543555
011 44 1527 541380
FAX: 011 44 1527 546956
e-mail: hooks@partridge-of-redditch.com

Tiemco (TMC)
Umpqua Feather Merchants
PO Box 700
Glide, OR 97443
(800) 322-3218
(541) 496-3512
FAX: (541) 496-0150

Turrall
Turrall USA
45 Knobhill Rd.
Sharon, MA 02067
(781) 784-3732
FAX: (781) 784-3732
e-mail: SPranTural@aol.com

H. Turrall & Co.
Dolton, Winkleigh
Devon, EX19 8QJ
England
011 441 805 804 352
FAX: 011 805 804 572

Varivas
True World Outdoor (Retail & Wholesale)
129 Industrial Ave.
Little Ferry, NJ 07643
(201) 451-8800
1-877-827-4827
FAX: (201) 451-9106
E-mail: sales@varivas.com
Web: www.varivas.com

VMC
VMC Inc.
Suite #10
1901 Oakcrest Ave.
St. Paul, MN 55113
(612) 636-9649
FAX: (612) 636-7053
Web: www.vmchooks.com

VMC Pêche
12, rue du Général de Gaulle
90120 Morvillars France
(33) 84 27 89 10
FAX: (33) 84 23 50

REFERENCES

Dan Bailey's Fly Shop. Catalog. Livingston, MT: Dan Bailey's Fly Shop, 1996.

Daiichi Fly Tyer's Guide. Brochure. Elba, NY: Angler Sport Group.

Angler Sport Group 1999 Product Catalog. Elba, NY: Angler Sport Group.

Eagle Claw 1999 Profit Book. Catalog no. 999P. Denver: Wright McGill Co., 1998.

1996 Eagle Claw. Catalog. Denver: Wright and McGill Co., 1995.

Eagle Claw Lazer Sharp Fly Hooks. Brochure. Distributed by Gudebrod.

Fina Comfort Fishing Gear, '99 Lure & Fly Products Guide. Catalog. Hyogo, Japan: Fina.

Gamakatsu Catch What You've Been Missing Catalog, 1995–96. Fife, WA: Gamakatsu USA.

Gamakatsu Fishing Ain't Luck Consumer Catalog.

Guide Book TMC Fly Hooks. Brochure. Umpqua Feather Merchants.

Introducing the Amazing New Vanadium Fly Hooks. Brochure. St. Paul, MN: VMC, n.d.

Varivas, Fly Fishing Products for the Serious Angler. Brochure. Little Ferry, NJ: True World Outdoor.

Hayabusa Fishing Hooks-Fly. Flyer. Japan 1993.

Kamasan, Hooks Around the World, 1997 Price List, Hooks and Tackle. Catalog. Kennington, Oxford, UK: Kamatsu Kamasan Ltd., 1997.

New from Mustad. Catalog 059810M13000078. Auburn, NY.

Mustad the Fish Hook Catalog. Catalog no. 0695.5M.

Mustad Advertising Slick. Leaflet.

Mustad, the Fish Hook Catalog. Catalog no. 0594.25.

Orvis Fly-Tying Materials and Rod Building 1996.

Owner 1999. Catalog. Costa Mesa, CA.

Partridge of Redditch fly hook leaflets.

Spirit River, Inc. Fly Tyer's Material Guide. Spirit River.

Sprite of Redditch; Gaelic Supreme leaflets.

The Tiemco Guide to Hooks. Brochure. Umpqua Feather Merchants.

Turrall. Catalog. H. Turrall & Co.

Turrall Fly-Tying. Catalog. H. Turrall & Co.

World's Finest Fish Hooks, VMC. Catalog. Printed in France.